Neurochemistry
of
Abused Drugs

Neurochemistry
of
Abused Drugs

Edited by
Steven B. Karch, MD, FFFLM

Consultant Pathologist and Toxicologist
Berkeley, California

CRC Press
Taylor & Francis Group
Boca Raton   London   New York

CRC Press is an imprint of the
Taylor & Francis Group, an informa business

CRC Press
Taylor & Francis Group
6000 Broken Sound Parkway NW, Suite 300
Boca Raton, FL 33487-2742

© 2008 by Taylor & Francis Group, LLC
CRC Press is an imprint of Taylor & Francis Group, an Informa business

---

Library of Congress Cataloging-in-Publication Data

Neurochemistry of abused drugs / [edited by] Steven B. Karch.
    p. ; cm.
  "A CRC title."
  Includes bibliographical references and index.
  ISBN-13: 978-1-4200-5441-5 (hardcover : alk. paper)
  ISBN-10: 1-4200-5441-4 (hardcover : alk. paper)
  1. Drugs of abuse--Pathophysiology. 2. Drugs of abuse--Physiological effect. 3. Neurochemistry. 4. Neurotoxicology. I. Karch, Steven B.
  [DNLM: 1. Substance-Related Disorders--physiopathology. 2. Brain--drug effects. 3. Neurotoxicity Syndromes--etiology. 4. Substance-Related Disorders--complications. WM 270 N4943 2007] I. Title.

  Q11.N4889 2007
  616.8'047--dc22   2007008113

Visit the Taylor & Francis Web site at
http://www.taylorandfrancis.com

and the CRC Press Web site at
http://www.crcpress.com

Contents

Preface

The first reports of neurological disease complicating drug abuse were published almost as soon as purified cocaine and morphine became abundant and cheap in the late 1800s. Today, neurological complaints are among the most common manifestations of drug abuse. At the molecular level, experimental studies have provided some surprising insights into the effects of drug abuse on the brain and plausible explanations for some types of drug toxicity. For example, evidence is emerging that nitric oxide formation plays an important role in cocaine neurotoxicity. Mice sensitized to cocaine administration initially tolerated doses of cocaine that became lethal after less than a week, but pretreatment with agents that inhibit nitric oxide synthetase completely abolished the sensitization process, and all test animals survived. Whether similar changes occur in humans remains to be determined.

All abused drugs, not just cocaine, activate immediate-early gene expression in the striatum, although different drugs induce somewhat different changes. Most activate immediate-early gene expression in several regions of the forebrain, including portions of the extended amygdala, lateral septum, midline/intralaminar thalamic nuclei, and even the cerebral cortex. These changes are especially striking in the case of cocaine. Postmortem studies have shown that, in humans, the numbers of both D1 and D2 dopamine receptors are altered by cocaine use, even with relatively low doses of cocaine. Strong evidence suggests that alterations in dopamine transmitters and receptors play a key role in the process of cocaine addiction and toxicity, but clearly much more is involved.

It has always been a puzzling question that the neurotoxic changes produced by some amphetamines share a strong resemblance with those seen in some degenerative disorders. The answer is no longer quite so puzzling. They share a number of common targets, including the ubiquitin–proteasome system, and both the ubiquitin–proteasome pathway and beta–arrestin are molecular targets of neurotoxicity. This knowledge may very well result in treatments for both.

Even though the mu receptor was first cloned nearly two decades ago, opiate addiction remains a major public health concern. However, the molecular mechanisms of opiate addiction are slowly becoming understood. Many of the changes that occur in neurons exposed to morphine have been known for some time, but not that much is known about the changes in gene expression that underlie these effects. With the advent of microarray analysis and quantitative (real time) PCR, it is now possible to examine the gene expression changes that occur during morphine withdrawal. The possibility of safely and effectively treating addicts (and relieving pain) is a tempting target and will, no doubt, occur in the near future.

The chapters of this book describe the Pandora's box of addictions that now face our society — cocaine, tobacco, methamphetamine, and MDMA. More importantly, they describe what is know at this moment about the neurochemical substrates underlying these disorders. Progress in molecular biology will be stunted until scientists understand the clinical presentations of the diseases they are trying to characterize. Clinicians stand little chance of curing addiction until they understand the underlying neurochemistry. One might say that this volume contains something for everybody.

The Editor

Steven B. Karch, M.D., FFFLM, received his undergraduate degree from Brown University. He attended graduate school in anatomy and cell biology at Stanford University. He received his medical degree from Tulane University School of Medicine. Dr. Karch did postgraduate training in neuropathology at the Royal London Hospital and in cardiac pathology at Stanford University. For many years he was a consultant cardiac pathologist to San Francisco's Chief Medical Examiner.

In the U.K., Dr. Karch served as a consultant to the Crown and helped prepare the cases against serial murderer Dr. Harold Shipman, who was subsequently convicted of murdering 248 of his patients. He has testified on drug abuse–related matters in courts around the world. He has a special interest in cases of alleged euthanasia, and in episodes where mothers are accused of murdering their children by the transference of drugs, either *in utero* or by breast feeding.

Dr. Karch is the author of nearly 100 papers and book chapters, most of which are concerned with the effects of drug abuse on the heart. He has published seven books. He is currently completing the fourth edition of *Pathology of Drug Abuse*, a widely used textbook. He is also working on a popular history of Napoleon and his doctors.

Dr. Karch is forensic science editor for Humana Press, and he serves on the editorial boards of the *Journal of Cardiovascular Toxicology*, the *Journal of Clinical Forensic Medicine* (London), *Forensic Science, Medicine and Pathology*, and *Clarke's Analysis of Drugs and Poisons*.

Dr. Karch was elected a fellow of the Faculty of Legal and Forensic Medicine, Royal College of Physicians (London) in 2006. He is also a fellow of the American Academy of Forensic Sciences, the Society of Forensic Toxicologists (SOFT), the National Association of Medical Examiners (NAME), the Royal Society of Medicine in London, and the Forensic Science Society of the U.K. He is a member of The International Association of Forensic Toxicologists (TIAFT).

Contributors

Michael H. Baumann, Ph.D.
Clinical Psychopharmacology Section
Intramural Research Program
National Institute on Drug Abuse
National Institutes of Health
Department of Health and Human Services
Baltimore, Maryland

John W. Boja, Ph.D.
U.S. Consumer Product Safety Commission
Directorate for Health Sciences
Bethesda, Maryland

Darlene H. Brunzell, Ph.D.
Department of Psychiatry
Yale University School of Medicine
New Haven, Connecticut

Kelly P. Cosgrove, Ph.D.
Department of Psychiatry
Yale University School of Medicine
New Haven, Connecticut
and
VA Connecticut Healthcare System
West Haven, Connecticut

Irina Esterlis, Ph.D.
Department of Psychiatry
Yale University School of Medicine
New Haven, Connecticut
and
VA Connecticut Healthcare System
West Haven, Connecticut

Colin N. Haile, Ph.D.
Department of Psychiatry
Yale University School of Medicine
New Haven, Connecticut
and
VA Connecticut Healthcare System
West Haven, Connecticut

Suchitra Krishnan-Sarin, Ph.D.
Department of Psychiatry
Yale University School of Medicine
New Haven, Connecticut
and
VA Connecticut Healthcare System
West Haven, Connecticut

Deborah C. Mash, Ph.D.
Departments of Neurology and Molecular
  and Cellular Pharmacology
University of Miami
Miller School of Medicine
Miami, Florida

William M. Meil, Ph.D.
Department of Psychology
Indiana University of Pennsylvania
Indiana, Pennsylvania

Richard B. Rothman, M.D., Ph.D.
Clinical Psychopharmacology Section
Intramural Research Program
National Institute on Drug Abuse
National Institutes of Health
Department of Health and Human Services
Baltimore, Maryland

Julie K. Staley, Ph.D.
Department of Psychiatry
Yale University School of Medicine
New Haven, Connecticut
and
VA Connecticut Healthcare System
West Haven, Connecticut

CHAPTER 1

The Dopamine Transporter and Addiction

William M. Meil, Ph.D.[1] **and John W. Boja, Ph.D.**[2]

[1] Department of Psychology, Indiana University of Pennsylvania, Indiana, Pennsylvania
[2] U.S. Consumer Product Safety Commission, Directorate for Health Sciences, Bethesda, Maryland

CONTENTS

Dopamine transporter (DAT) is a distinctive feature of dopaminergic neurons, discovered more than 20 years ago.[1–5] DAT is the major mechanism for the removal of released dopamine (DA). DA is actively transported back into dopaminergic neurons via a sodium- and energy-dependent mechanism.[6–8] Like other uptake carriers, DAT is regulated by a number of drugs including cocaine, amphetamine, some opiates, and ethanol. It is this interaction with DAT and the resulting increase in synaptic DA levels that have been suggested to be the basis for the action of several drugs of abuse. The dopaminergic hypothesis of drug abuse has been proposed by a number of researchers.[9,10] Di Chiara and Imperato[11] observed the effects of several drugs of abuse on DA levels in the nucleus accumbens and caudate nucleus using microdialysis. Drugs such as cocaine, amphetamine, ethanol, nicotine, and morphine were all observed to produce an increase in DA, especially in the nucleus accumbens. Drugs that are generally not abused by humans, such as bremazocine, imipramine, diphenhydramine, or haloperidol, decreased DA or increased DA in the caudate nucleus only. It was, therefore, concluded[11] that drugs abused by humans preferentially increase brain DA levels in the nucleus accumbens, whereas psychoactive drugs not abused by humans do not. By employing this hypothesis of drug reward as a starting point, this chapter reviews evidence regarding the function of DAT and the interaction of several drugs of abuse on DAT.

1.1 DOPAMINE UPTAKE

The uptake of DA depends on a number of factors,[4–6,12–15] including temperature, sodium,[16–19] potassium,[6,16] and chloride,[7,20] but not calcium.[6] Krueger[21] suggested that dopamine transport occurred by means of two sodium ions and one chloride ion carrying a net positive charge into the neuron, which is utilized to drive DA against its electrochemical gradient. More recently, McElvain and Schenk[22] proposed a multisubstrate model of DA transport. In this model it was proposed that either one molecule of DA or two sodium ions bind to DAT in a partially random mechanism. Chloride binds next and it is only then that the DAT translocates from the outside of the neuron to the inside (Figure 1.1). Cocaine inhibition of DA transport occurs with cocaine binding to the sodium-binding site and changing the conformation of the chloride-binding site, thus preventing the binding of either and ultimately inhibiting dopamine uptake. DA uptake by cocaine appeared to be uncompetitive inhibition, whereas the binding of sodium and chloride are competitively inhibited. This action is present only with neuronal membrane-bound DAT because cocaine does not appear to inhibit the reuptake of DA to the vesicles via the vesicular transporter.[23] Moreover, site-directed mutations of DAT hydrophobic regions[24] or the carboxyl-terminal tail[25] have resulted in differential effects on cocaine analogue binding and dopamine uptake.

A recent review of the literature on the amino acid structure of DAT stated that uptake of dopamine is dependent on multiple functional groups of amino acids within DAT.[26] The authors

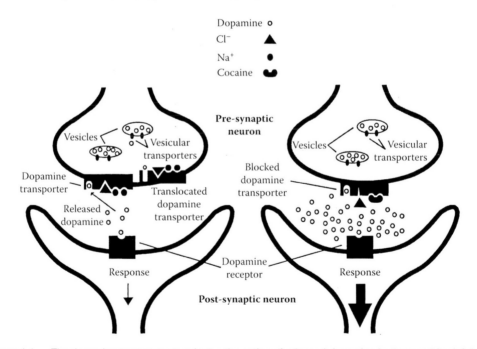

Figure 1.1   The dopamine transporter terminates the action of released dopamine by transport back into the presynaptic neuron. Dopamine transport occurs with the binding of one molecule of dopamine, one chloride ion, and two sodium ions to the transporter; the transporter then translocates from the outside of the neuronal membrane into the inside of the neuron.[22] Cocaine appears to bind to the sodium ion binding site. This changes the conformation of the chloride ion binding site; thus dopamine transport does not occur. This blockade of dopamine transport potentiates dopaminergic neurotransmission and may be the basis for the rewarding effects of cocaine.

suggested that the amino acid functional groups of Phe[69], Phe[105], Phe[114], Phe[155], Thr[285], Phe[319], Phe[311], Pro[394], Phe[410], Ser[527], Phe[520], Tyr[533], and Ser[538] in rat DAT and Val[55] and Ser[528] in human DAT appear to be involved in DAT uptake.

1.2 ABUSED DRUGS AND THE DOPAMINE TRANSPORTER

1.2.1 Cocaine

Cocaine has several mechanisms of action: inhibition of DA, norepinephrine, and serotonin reuptake, as well as a local anesthetic effect. While the stimulating and reinforcing effects of cocaine have been recognized for quite some time, it was not until recently that the mechanism for these effects was elucidated. The stimulatory effects of cocaine were first associated with the ability of cocaine to inhibit the reuptake of DA.[27,28] Saturable and specific binding sites for [³H]cocaine were then discovered by Reith using whole mouse brain homogenates.[29] When striatal tissue was utilized as the sole tissue source, Kennedy and Hanbauer[30] were able to correlate the pharmacology of [³H]cocaine binding and [³H]DA uptake inhibition and, thereby, hypothesized that the binding site for [³H]cocaine was in fact DAT. By using the data from binding experiments, it has been possible to correlate the strong reinforcing properties of cocaine with blockade of DAT rather than inhibition of either the serotonin (SERT) or norepinephrine transporters (NET).[31,32]

By using radiolabeled cocaine[33-35] or analogues of cocaine such as WIN 35,065-2,[30] WIN 35,428,[33,34] RTI-55,[35-43] and RTI-121,[44,45] it is possible to visualize the distribution of these drugs within the brain; the pattern of binding demonstrated by cocaine and its analogues appears to coincide with the distribution of dopamine within the brain. Areas of the brain with the greatest amount of dopaminergic innervation, such as the caudate, putamen, and nucleus accumbens, also demonstrate the greatest amount of binding, whereas moderate amounts of binding are observed in the substantia nigra and ventral tegmental areas. Recently specific antibodies to the DAT have been developed.[46] Visualization of the distribution of DAT within the brain using these antibodies demonstrated that there was a good correlation with cocaine binding.

Several unrelated compounds have been demonstrated to bind to the DAT, such as [³H]mazindol,[47] [³H]nomifensine,[48] and [³H]GBR 12935.[49] However, while these compounds also inhibit the reuptake of DA, they do not share the powerful reinforcing properties of cocaine. The question of why these compounds are non-addictive while cocaine is quite addictive remains unanswered. Several possibilities exist: Schoemaker et al.[50] observed that [³H]cocaine binds to both a high- and low-affinity site on the DAT, whereas other ligands such as [³H]mazindol,[47] [³H]nomifensine,[48] and [³H]GBR 12935[49] bind solely to a single high-affinity site. This does not indicate that the two binding sites demonstrated by cocaine and its analogues[43,44,51-54] represent two distinct sites, however, because both the high- and low-affinity sites arise from a single expressed cDNA for the DAT.[55] Another difference may be the pattern of binding, in that [³H]mazindol binds to different sites in the brain than those observed for [³H]cocaine.[56] In addition, the rate of entry into the brain is different for these different compounds. Mazindol and GBR 12935 have been demonstrated to enter the brain and occupy receptors much more slowly than cocaine.[57,58] At the present time it is still unclear which of these or other possible factors promote the strong reinforcing properties of cocaine.

Recently, mice lacking the gene for DAT have been developed;[59] DA is present in the dopaminergic extracellular space of the homozygous mice almost 100 times longer than it is present in the normal mouse. The homozygous mice were hyperactive compared to normal mice and, as expected, cocaine did not produce any effect in the locomotor activity of the homozygous mice. These results provide further evidence to support the concept of the DAT as a cocaine receptor. However, mice lacking DAT do show cocaine reinforcement.[60-63] Possible explanations for this observation include a role of SERT[60,62] or NET in the psychoactive effects of cocaine.

1.2.2   Amphetamine

Amphetamine and its analogues, including but not limited to methamphetamine, methylene-dioxyamphetamine (MDA), and methylenedioxymethamphetamine (MDMA), increase brain DA levels.[64–76] Amphetamine has been postulated to increase brain DA levels either by increasing DA release or by blocking DA reuptake. Hadfield[77] observed amphetamine blockade of DA reuptake; however, reuptake inhibition occurred only at doses of amphetamine (ED_{50} = 65 mg/kg) that were much higher than the doses observed to increase release. While reuptake blockade may play a role in the ability of amphetamine to elevate DA, blockade occurs only at doses near those that produce stereotypy or toxicity. On the other hand, amphetamine-stimulated DA release occurs at much lower doses. Amphetamine-stimulated DA release has been postulated to occur by two mechanisms: one involves the interaction of amphetamine with the DAT, which then produces a reversal of the DAT so that DA is transported out of neuron while amphetamine is transported out of the neuron.[77–85] The other proposes passive diffusion of amphetamine-mediated alteration of vesicular pH.[84] Using human DAT-transfected EM4 cells, Kahlig[86] observed both a fast and slow efflux of dopamine following amphetamine stimulation suggesting that amphetamine releases DA via the DAT in a quantum-like manner resulting in a slow DA release and in a faster channel-like manner.

Besides this purported action on DAT, amphetamine has also been suggested to act upon the vesicular transporter as well. Pifl et al.[87] examined COS cells transfected with cDNA for either DAT or the vesicular transporter, or both. A marked increase in DA release was noted in cells that expressed both DAT and the vesicular transporter when compared to the release from cells that express only DAT or the vesicular transporter. The mechanism of action for amphetamine was further defined with the work of Giros et al.[59] In transgenic mice lacking the DAT, amphetamine did not produce hyperlocomotion or release DA.

In summary, the DAT appears to be the primary site of action for amphetamine-induced DA release via its activity on the DAT because amphetamine appears to employ DAT to transport DA out of the neuron while, at the same time, amphetamine may be sequestered in the neuron. The sequestered amphetamine then may release vesicular DA by altering vesicular pH or via interactions with the vesicular transporter.

1.2.3   Opiates

Opiate drugs share the ability to elevate extracellular DA concentrations in the nucleus accumbens,[88–90] possibly implicating mesolimbic DA activity in the abuse liability of these compounds. Whereas the locomotor[91] and reinforcing effects[92,93] of opiates may occur through DA-independent pathways, there is also evidence for dopaminergic mediation of these effects.[94,95] Lesions of dopaminergic neurons[96,97] or neuroleptic blockade of DA receptors[98,99] attenuate opiate reward as measured by intracranial electrical self-stimulation, conditioned place preference, and intravenous self-administration. In contrast to cocaine's ability to augment DA concentrations through direct action at DAT,[100] opiates appear to enhance DA concentrations primarily by indirectly stimulating DA neurons.[101,102]

However, evidence suggests that some opiates also act at DAT. Das et al.[103] reported that U50-488H, a synthetic κ-opiate agonist, and dynorphin A, an endogenous κ ligand, dose-dependently inhibit [^{3}H]DA uptake in synaptosomal preparations from the rat striatum and nucleus accumbens. Inhibition of [^{3}H]DA uptake by U50-488H was not reversed by pretreatment with the opiate antagonists naloxone and nor-binaltorphine, suggesting that this effect is mediated through direct action at DAT rather than an indirect effect at κ receptors. However, the effects of another κ-opiate agonist, U69593, do not appear mediated by the DAT since U69593 failed to attenuate GBR 12909- and WIN 35,428-induced cocaine seeking behavior.[104]

Meperidine, an atypical opiate receptor agonist with cocaine-like effects, has been shown to act at the DAT.[105] Meperidine inhibited [^{3}H]DA uptake in rat caudate putamen with a maximal effect less than that achieved with cocaine. This suggests that meperidine may predominantly act at the high-affinity transporter site. Meperidine also displaced [³H]WIN 35,428 binding in a manner consistent with a single site affinity. Because meperidine shares key structural features with the phenyltropane analogues of cocaine, it is possible that these common structural features account for the cocaine-like actions of meperidine rather than any characteristics intrinsic to opiates. Similarly, fentanyl, a μ-opiate agonist structurally related to meperidine, decreased [¹²³I]β-CIT binding in the basal ganglia of a single human subject and in rats, supporting the direct action of some opiates on dopamine reuptake.[106] In contrast, selective μ and opiate agonists failed to inhibit [³H]DA uptake in the striatum and nucleus accumbens across the same range of doses. Morphine, a μ-opiate agonist, also did not inhibit [³H]DA uptake or displace [³H]WIN 35,428 binding in the striatum[105] or displace [³H]GBR 12935 binding in basal forebrain.[107] Conditioned place preference to morphine is increased in DAT knockout mice.[108]

Although opiates and psychostimulants may possess different sites of action, it has been suggested that cross-sensitization of their addictive properties may result from overlapping neural targets. Examining the localization of κ-opioid receptor and DAT antisera in nucleus accumbens shell of the rat, κ-opioid receptor labeling was seen primarily in axon terminals and DAT labeling was observed exclusively in axon terminals. Thus, opiate agonists in the nucleus accumbens shell may modulate DA release primarily via control of presynaptic neurotransmitter secretion that may influence or be influenced by intracellular DA.[109]

Although morphine appears to lack direct action at DAT, research suggests that chronic morphine may alter DAT expression. Repeated, but not acute, administration of morphine to rats decreased the B_{max} of [³H]GBR 12935 binding in the anterior basal forebrain, including the nucleus accumbens, but not the striatum.[107] However, radioligand affinity was not different in either brain region. Neither acute nor chronic morphine administration inhibited binding at the serotonin transporter in the striatum or anterior basal forebrain, suggesting that transporter down-regulation was selective for brain regions important for the reinforcing and/or motivational properties of opiates. Because daily cocaine administration in rats also attenuates DA uptake in the nucleus accumbens and not the striatum,[110] chronic elevation of DA release and a subsequent reduction in DAT expression within the nucleus accumbens may prove important in the development of drug addiction. The effects of chronic morphine administration on DAT activity may also be related to withdrawal status of the animal. Rats implanted with morphine pellets for 7 days and examined with the pellets intact showed [³H]GBR 12935 binding was increased in the hypothalamus and decreased in the striatum. Rats examined 16 h after removal of the pellets showed increased binding in both the hypothalamus and hippocampus.[111] However, recent research has demonstrated that twice daily escalating doses of morphine for 7 days altered mRNA levels for several dopamine receptors (D_2R and D_3R) but not the DAT in discrete regions of the rat brain.[112] Also, post-mortem examination of the striatum of nine chronic heroin users revealed modest reductions in measures of dopamine function but levels of vesicular monoamine transporters were comparable to controls.[113]

1.2.4 Phencyclidine

Both systemic and local infusions of phencyclidine (PCP) enhance extracellular DA concentrations in the nucleus accumbens[114,115] and prefrontal cortex.[116] PCP-induced elevations of extracellular DA concentrations may result from both indirect and direct effects on the dopaminergic system. NMDA receptors exert a tonic inhibitory effect upon basal DA release in the prefrontal cortex[116,117] and in the nucleus accumbens through inhibitory effects on midbrain DA neurons.[118–120] Thus, PCP antagonism of NMDA receptors[121] may facilitate DA release by decreasing the inhibition of central dopaminergic activity.

PCP also increases calcium-independent [³H]DA release from dissociated rat mesencephalon cell cultures[122] and striatal synaptosomes.[123] PCP has been found to be a potent inhibitor of [³H]DA uptake in rat striatum,[124–127] to competitively inhibit binding of [³H]BTCP, a PCP derivative and potent DA uptake inhibitor in rat striatal membranes,[128] and to inhibit [³H]cocaine binding.[129] In addition, (trans)-4-PPC, a major metabolite of PCP in humans,[130] inhibits [³H]DA uptake in rat striatal synaptosomes with comparable potency to PCP and thus it may be involved in the psychotomimetic effects of PCP.[124] Recently it was reported that PCP exerts some direct actions at the DAT in the primate striatum using positron emission tomography. Moreover, it was suggested that GABA may also modulate PCP-induced augmentation of DA in the primate striatum.[131]

Despite the profound effect PCP exerts on mesolimbic DA activity, evidence suggests that the reinforcing properties of PCP are not dopamine dependent. Carlezon and Wise[132] have reported that rats will self-administer PCP into the ventromedial region of the nucleus accumbens, as well as NMDA receptor antagonists that do not inhibit DA reuptake. Co-infusion of the DA antagonist sulpiride into the nucleus accumbens inhibits intracranial self-administration of nomifensine, but not PCP. Moreover, rats self-administer PCP into the prefrontal cortex, an area that will not maintain self-administration of nomifensine.[133] Therefore, the reinforcing effects of PCP in the nucleus accumbens and prefrontal cortex appear to be related to PCP blockade of NMDA receptor function rather than its dopaminergic actions. Instead, PCP-induced elevations of extracellular DA may mediate other behavioral effects of PCP, such as its stimulant effects on locomotor activity.[134] The differential effects on locomotor activation of PCP and cocaine do not appear mediated though direct action at the DAT.[135]

1.2.5   Marijuana

Recent progress has greatly expanded our knowledge of the endocannabinoid system and the ways in which Δ^9-tetrahydrocannabinol (Δ^9-THC), the primary psychoactive component of marijuana, acts upon this system. Advances have included the identification of central cannabinoid receptors (CB_1) as abundant primarily presynaptic G protein–coupled receptors sensitive to endogenous transmitters (anandamide, 2-AG) that function as retrograde transmitters and alter presynaptic neurotransmitter release.[136] The identification of synthetic ligands that act as agonists and antagonists at the CB_1 receptor has also greatly furthered our understanding of the endocannabinoid system and the effects of Δ^9-THC in the brain.[137]

Activation of dopaminergic circuits known to play a pivotal role in mediating the reinforcing effects of other abused drugs also results from cannabinoid administration.[138] Systemic or local injections of Δ^9-THC enhance extracellular dopamine concentrations in the rat prefrontal cortex,[139,140] caudate,[141] nucleus accumbens,[142,143] and ventral tegmental.[144,145] In addition, Δ^9-THC augments both brain stimulation of reward and extracellular DA concentrations in the nucleus accumbens in Lewis rats, linking dopaminergic activity with the rewarding properties of marijuana.[143]

Recent research is beginning to define the interactions between DA and endocannabinoids in regions critical for our understanding of the reinforcing effects of Δ^9-THC. Activity-dependent release of endocannabinoids from the ventral tegmental area appears to serve as a regulatory feedback mechanism to inhibit synaptic inputs in response to DA neuron bursting and thus regulating firing patterns that may fine-tune DA release from afferent terminals.[146] Similarly, DA neurons in the prefrontal cortex have been suggested to release endocannabinoids to shape afferent activity and ultimately their own behavior.[147] Research has also begun to shed light on the intracellular signaling pathways activated by THC. Acute administration of Δ^9-THC produces phosphorylation of the mitogen-activated protein kinase/intracellular signal-regulated kinase (MAP/ERK) in the dorsal striatum and nucleus accumbens. This activation, corresponding to both neuronal cell bodies and the surrounding neuropil, is blocked by pretreatment with DA D_1, and to a lesser extent DA D_2 and NMDA glutamate, antagonists.[148] Given that ERK inhibition was found to block conditioned place preference for Δ^9-THC, these findings suggest dopaminergic influence of Δ^9-THC intracellular effects is important for the rewarding effects of Δ^9-THC.[148]

Facilitation of dopaminergic activity by Δ^9-THC may result from multiple mechanisms. Δ^9-THC increases DA synthesis[149] and release[150] in synaptosomal preparations. In addition, using *in vivo* techniques, Δ^9-THC has been reported to augment potassium-evoked DA release in the caudate[141] and increase calcium-dependent DA efflux in the nucleus accumbens.[142] However, whereas Δ^9-THC produces a dose-dependent augmentation of somatodendritic DA release in the ventral tegmental area, it fails to simultaneously alter accumbal DA concentrations.[144] Because local infusions of Δ^9-THC through a microdialysis probe did elevate nucleus accumbens DA concentrations, modulation of DA activity in the nucleus accumbens is likely to result from presynaptic effects.

Δ^9-THC also acts directly at the DAT to affect DA uptake. At low concentrations Δ^9-THC stimulates uptake of [³H]DA in synaptosomal preparations of rat brain striatum and hypothalamus.[150] Similarly, mice injected with Δ^9-THC showed increased [³H]DA uptake into striatal synaptosomes and, to a greater extent, in cortical synaptosomes.[151] At higher concentrations Δ^9-THC inhibits uptake of [³H]DA in rat striatal[150,152,153] and hypothalamic[150] synaptosomes. Also consistent with the hypothesis that Δ^9-THC blocks DA uptake, using *in vivo* electrochemical techniques, it has been reported that Δ^9-THC and the DA-reuptake blocker nomifensine produce identical augmentation of voltammetric signals corresponding to extracellular DA.[141] While Δ^9-THC has a similar biphasic effect on norepinephrine uptake in hypothalamic and striatal synaptosomes[150] and increases uptake of 5-HT and GABA in cortical synaptosomes,[151] the psychoactive effects of Δ^9-THC are most likely related to dopaminergic activity because less potent and nonpsychoactive THC derivatives show much less effect on DA uptake than does Δ^9-THC.[151] It is only recently that the effects of Δ^9-THC exposure on human DAT levels have been examined and while it appears that postmortem DAT levels in the caudate of individuals with schizophrenia may be influenced by Δ^9-THC, this result may be of limited generalizability given that people suffering schizophrenia tend to show reduced DAT levels regardless of history of THC use.[154]

Δ^9-THC clearly has profound effects on dopaminergic activity in areas important to the maintenance of the reinforcing effects of other abused compounds. Research relating the persistence of Δ^9-THC-induced ventral tegmental DA neuron firing in animals chronically treated with Δ^9-THC to the lack of tolerance to marijuana's euphoric effects further bolsters this link.[155] The ability of Δ^9-THC to facilitate intracranial electrical self-stimulation in the median forebrain bundle has long been established,[156] however, only recently have the reinforcing effects of Δ^9-THC been clearly demonstrated using conditioned place preference[148,157] and drug self-administration[157,158] procedures. With advances in our understanding of the endocannabinoid system and the further establishment of animal models of Δ^9-THC-induced reinforcement, increased understanding of marijuana's abuse liability can be expected in coming years. The observation that CB₁ receptor antagonism attenuates the reinstatement of heroin self-administration has also implicated the endocannabinoid system in the mechanisms underlying addiction and suggests a potential therapeutic niche for cannabinoid ligands.[159]

1.2.6 Ethanol

Ethanol also alters the dopaminergic system. Administration of ethanol has been shown to release DA *in vivo*[160–162] and *in vitro*.[163–171] The mechanism(s) by which ethanol increases brain DA levels are slowly beginning to be understood and may involve modulation of DAT activity. Tan et al.[172] examined [³H]DA uptake in brain synaptosomes prepared from rats in various stages of intoxication. [³H]DA uptake was inhibited by ethanol for as long as 16 h following the withdrawal of ethanol. A potential mechanism by which ethanol might work to increase DAT function may involve regulation of DAT expression on the cell surface as [³H]DA has been shown to accumulate following ethanol administration in human DAT expressing *Xenopus* oocytes in parallel with cell surface DAT binding measured by [³H]WIN 35,428.[173] Moreover, sites on the second intracellular loop of the DAT have been identified that appear important for ethanol modulation of DAT activity.[174] However, further research on the effects of ethanol on DAT function is needed given that recent research suggests acute ethanol attenuates DAT function in rat dorsal striatum and ventral striatum of anesthetized rats and tissue suspensions.[175]

Ethanol also increased both spontaneous release and Ca^{2+}-stimulated release of DA, but decreased the amount of K^+-stimulated released DA in rat striatum.[160,172] The increased amount of DA release is not due to nonspecific disruption of the neuronal membrane because acetylcholine levels are not altered.[162] Thus, it appears that ethanol can affect both the release and reuptake of DA via a specific mechanism. However, research investigating ethanol-induced DA release in rat nucleus accumbens slices suggests the mechanism is different from that underlying the effects of depolarization with electrical stimulation or high potassium levels and implicate nonexocytotic mechanisms.[177] Using no net flux microdialysis methodology to examine the effects of intraperitoneal injections of ethanol-induced increases in DA in the rat nucleus accumbens, it was suggested the primary mechanism by which ethanol augments extracellular DA levels is by facilitating release from terminals rather than by blocking the DAT.[178] However, research showing attenuated ethanol preference and consumption in female DAT knockout mice suggests ethanol's action on DAT may be relevant to ethanol-induced reward.[179]

A transesterification product of ethanol and cocaine has been discovered. Benzoylecgonine ethyl ester or cocaethylene was first described by Hearn et al.[180] Cocaethylene possessed similar affinity for the DAT as cocaine and also inhibited DA uptake[180–183] and increased *in vivo* DA levels.[184,185] Cocaethylene has lower affinity for the serotonin transporter than cocaine. Cocaethylene produces greater lethality in rats, mice, and dogs than cocaine[186–189] and may potentiate the cardiotoxic effects and tendency toward violence from cocaine or alcohol in humans.[190] While showing a similar pharmacological and behavioral profile as cocaine, cocaethylene appears less potent than cocaine in human subjects.[191] Anecdotal reports from human addicts and experimental results with animal subjects support the hypothesis that alcohol is often ingested with cocaine in order to attenuate the negative aftereffects of cocaine.[192]

1.2.7   Nicotine

Nicotine increased DA levels both *in vivo*[11,193] and *in vitro*.[194–196] Nicotine[197] and its metabolites[198] were found to both release and inhibit the reuptake of DA in rat brain slices, with uptake inhibition occurring at a lower concentration than that required for DA release. In addition, the (−) isomer was more potent than the (+) isomer.[197] However, the effects of nicotine upon DA release and uptake were only apparent when brain slices were utilized because nicotine was unable to affect DA when a synaptosomal preparation was utilized.[197] These results indicate that nicotine exerts its effects upon the DAT indirectly, most likely via nicotine acetylcholine receptors. This finding was supported by the results of Yamashita et al.[199] in which the effect of nicotine on DA uptake was examined in PC12 and COS cells transfected with rat DAT cDNA. Nicotine inhibited DA uptake in PC12 cells that possess a nicotine acetylcholine receptor. This effect was blocked by the nicotinic antagonists hexamethonium and mecamylamine. Additionally, nicotine did not influence DA uptake in COS cells, which lack nicotinic acetylcholine receptors.

Interestingly, a series of cocaine analogues that potently inhibited cocaine binding also inhibited [³H]nicotine and [³H]mecamylamine binding.[200] It was concluded that the inhibition by these cocaine analogues involves its action on an ion channel on nicotinic acetylcholine receptors. Recently several studies have further investigated the ability of nicotine to regulate DAT function. In slices from rat prefrontal cortex, but not the striatum or nucleus accumbens, nicotine enhances amphetamine-stimulated [³H]DA release via the DAT. Moreover, the nicotinic acetylcholine receptors responsible for mediating amphetamine-induced [³H]DA release in the prefrontal cortex were found to be at least partially localized on nerve terminals.[201,202] However, nicotine was found to augment DA clearance in the striatum and prefrontal cortex in a mecamylamine-sensitive manner, suggesting nicotinic acetylcholine receptors also modulate striatal DAT function.[203] Chronic nicotine and passive cigarette smoke exposure increase DAT mRNA in the ventral tegmental area in the rat[204] and other data suggest that changes in DAT numbers following repeated nicotine exposure may be behaviorally relevant since increases in DAT and D_3 receptors in the nucleus accumbens appear to be at least partially responsible for gender differences in behavioral sensitization to nicotine.[205]

1.3 ABUSED DRUGS AND GENETIC POLYMORPHISM OF THE DOPAMINE TRANSPORTER

Familial, twin, and adoption studies suggest there may be a genetic predisposition toward drug addiction.[206] Genetic polymorphisms across several neurotransmitter systems, including the dopaminergic system, have been linked to the development of drug addiction.[207] In humans the DAT gene (DAT1) has a variable number of tandem repeats (VNTR) in the 3′-untranslated region known to influence gene expression.[208] Most research suggests the longer 10-repeat allele yields greater DAT1 expression than the 9-repeat allele.[209] According to the reward deficiency syndrome hypothesis alterations in various combinations of genes, including DAT1, may provide some individuals with an underactive reward system and increase the likelihood that they will seek stimulation from the environment including stimulation from abused drugs.[210]

Research has implicated DAT polymorphisms to numerous effects of addictive drugs and addictive liability. Cocaine users with the 9/9 and 9/10 genotypes appear more susceptible to cocaine-induced paranoia than those with the 10/10 genotype.[211] Recently Lott et al.[212] reported that healthy volunteers with the 9/9 genotype have a diminished responsiveness to acute amphetamine injections on measures of global drug effect, feeling high, dysphoria, anxiety, and euphoria. These results may be significant given a diminished response to alcohol has been linked to future development of alcoholism.[213] However, another study found no significant associations between DAT polymorphism and clinical variations in a population of methamphetamine abusers.[214] Genetic polymorphisms across opioid and monoaminergic systems have also been linked to the development of opiate addiction.[215] Genetic polymorphisms in both the SERT and the DAT were found to be related to opiate addiction.[216] Homozygosity at the serotonin transporter (especially 10/10) was related to the development of opiate addiction, whereas the genotype 12/10 appeared to be protective against opiate addiction. The DAT1, genotype 9/9 was associated with early opiate addiction. Opiate abuse under the age of 16 was also predicted by a combination of the serotonin transporter genotype 10/10 and the DAT1 genotype 10/10.[216] Studies have also begun to assess whether the risk of alcoholism may be mediated by genetic polymorphism in a variety of genetic targets, including the dopaminergic system, although conflicting results remain to be clarified. According to some research DAT polymorphism has not clearly been identified as a risk factor for the development of alcoholism,[217] but it has been associated with the development of severe alcohol withdrawal symptoms.[218] However, other research has suggested that DAT polymorphism is related to the development of alcoholism but not alcohol withdrawal.[219] The role of DAT polymorphism in nicotine addiction has received the most attention. Although there have been some conflicting reports,[220] most studies suggest the 9-repeated allele of the DAT is related to a decreased likelihood of being a smoker, a lower likelihood of smoking initiation prior to age 16, and longer periods of abstinence among smokers.[221–223] This latter finding is consistent with the reward deficiency syndrome hypothesis since individuals with the 9-repeated allele would be expected to have decreased DAT expression leading to higher levels of intracellular DA and therefore a reduced need for novelty and external reward including cigarettes. Clearly genetic polymorphism across a number of neurotransmitter systems plays a role in the development of drug addiction. However, several studies now implicate genetic variation at DAT as being a potential contributor to this mixture.

1.4 CONCLUSIONS

The dopaminergic system plays a role in the abuse liability for some, if not most, drugs. The stimulants — opiates, marijuana, nicotine, and ethanol — all interact directly or indirectly with

Table 1.1   Comparison of the Self-Administration of Various Drugs and the Effect That Drug Has on DAT

Drug	Self-Administered	Increases DA via DAT	Ref.
Cocaine	+	+	27, 28, 224
Amphetamine	+	+	78, 224
MDMA	+	+	73, 225
DMT	?	–	226
Mescaline	–	–	224, 227
LSD	–	–	224, 227, 228
Opiates	+	+	103, 224
Barbiturates	+	–	229
Benzodiazepines	+	?	229–232
Alcohol	+	+	172, 224
Caffeine	+	–	224, 233
Nicotine	+	(Indirect)	194–199, 234
Marijuana	+	+	150, 157, 158
PCP	+	+	124–127, 235

the dopaminergic system, and most of these have actions on the DAT (Table 1.1). Numerous lines of evidence suggest the positive reinforcement, or DA hypothesis, of addiction falls short in accounting for all aspects of addiction.[236] While many believe the elevation of DA within the mesolimbic DA system is a contributing factor to the abuse liability of drugs, considerable evidence supports the notion that neuroadaptive changes resulting from chronic drug use is what actually drives addictive behavior.[237] An understanding of the role of the DAT in the addictive process will likely involve the understanding of how drugs initially interact with the DAT as well as the effects of chronic drug exposure on DAT expression and function.

DAT occupancy alone does not impart a drug with addictive properties. Some drugs that interact with the DAT, such as cocaine, are quite addictive, while other drugs, such as mazindol, are not. There appears to be a temporal component in that, while mazindol interacts with the dopaminergic system, its entry into the brain is slow compared to that of cocaine.[56,57] The importance of the rate at which transporter occupancy occurs is also underscored by the observation that routes of drug administration, like smoking or intravenous injection, that lead to rapid entry into the brain, and for some drugs rapid DAT occupancy, are more likely to produce an intense "high" and have greater addictive potential than drug administration via oral or nasal routes, which are associated with delayed drug action in the brain.[238,239] In addition, baseline DA activity within the mesolimbic pathway may also be an important influence on psychostimulant-induced "high." The subjective high produced by methylphenidate appears related to both DAT occupancy and basal DA activity of the subject.[240,241] This result hints at the potential importance of genetic polymorphism within the dopaminergic system on addictive liability. Given that genetic polymorphism of the DAT has been tentatively linked to the addictive potential of several drugs, a better understanding of the contribution that genetic polymorphism of the DAT plays in the development of addiction will be valuable.

The cloning of the DAT[242,244] and its subsequent transfection into cells have allowed for the study of DAT in much greater detail. Moreover, the development of transgenic mice that lack DAT has now afforded the study of the mechanisms of action for many drugs.[59] Using these and other powerful new tools, in the future we may be better able to understand the role of DAT in the mechanisms of action for addictive drugs, the addictive process, and individual differences in a person's predisposition toward drug addiction.

REFERENCES

1. Glowinski, J. and Iversen, L. L., Regional studies of catecholamines in the rat brain. I. The disposition of [³H]norepinephrine, [³H]dopamine and [³H]dopa in various regions of the brain, *J. Neurochem.*, 13, 655, 1996.

2. Snyder, S.J. and Coyle, J.T., Regional differences in [³H]norepinephrine and [³H]dopamine uptake into rat brain homogenates, *J. Pharmacol. Exp. Ther.*, 165, 78, 1969.

3. Kuhar, M.J., Neurotransmitter uptake: a tool in identifying neurotransmitter-specific pathways, *Life Sci.*, 13, 1623, 1973.

4. Horn, A.S., Characteristics of neuronal dopamine uptake, in *Dopamine. Advances in Biochemical Psychopharmacology,* Roberts, P.J., Woodruff, G.N., and Iversen, L.L., Eds., Vol. 19, Raven Press, New York, 1978, 25.

5. Horn, A.S., Dopamine uptake: a review of progress in the latest decade, *Prog. Neurobiol.*, 34, 387, 1990.

6. Holz, K.W. and Coyle, J.T., The effects of various salts, temperature and the alkaloids veratridine and batrachotoxin on the uptake of [³H]-dopamine into synatosomes from rat striatum, *Mol. Pharmacol.,* 10, 746, 1974.

7. Kuhar, M.J. and Zarbin, M.A., Synaptosomal transport: a chloride dependence for choline, GABA, glycine, and several other compounds, *J. Neurochem.*, 30, 15, 1978.

8. Cao, C.J., Shamoo, A.E., and Eldefrawi, M.E., Cocaine-sensitive, ATP-dependent dopamine uptake in striatal synaptosomes, *Biochem. Pharmacol.*, 39, 49, 1990.

9. Koob, G.F. and Bloom, F.E., Cellular and molecular mechanisms of drug dependence, *Science*, 242, 715, 1988.

10. Kuhar, M.J., Ritz, M.C., and Boja, J.W., The dopamine hypothesis of the reinforcing properties of cocaine, *Trends Neurosci.,* 14, 299, 1991.

11. Di Chiara, G. and Imperato, A., Drugs abused by humans preferentially increase synaptic dopamine concentrations in the mesolimbic system of freely moving rats, *Proc. Natl. Acad. Sci. U.S.A.*, 85, 5274, 1988.

12. Coyle, J.T. and Snyder, S.H., Catecholamine uptake by synaptosomes in homogenates of rat brain: stereospecificity in different areas, *J. Pharmacol. Exp. Ther.*, 170, 221, 1969.

13. Iversen, L.L., Uptake processes for biogenic amines, in *Biochemistry of Biogenic Amines*, Vol. 3, Plenum Press, New York, 1975, 381.

14. Horn, A.S., Characteristics of transport in dopamine neurons, in *The Mechanism of Neuronal and Extraneuronal Transport of Catecholamines*, Paton, D.M., Ed., Raven Press, New York, 195, 1976.

15. Amara, S. and Kuhar, M.J., Neurotransmitter transporters: recent progress, *Annu. Rev. Neurosci.*, 16, 73, 1993.

16. Harris, J.E. and Baldessarini, R.J., The uptake of [³H]dopamine by homogenates of rat corpus striatum: effects of cations, *Life Sci.*, 13, 303, 1973.

17. Horn, A.S., *The Neurobiology of Dopamine*, Horn, A.S., Korf, J., and Westerink, B.H.C., Eds., Academic Press, New York, 1979, 217.

18. Zimanyi, I., Lajitha, A., and Reith, M.E.A., Comparison of characteristics of dopamine uptake and mazindol binding in mouse striatum, *Naunyn-Schmiedeberg's Arch. Pharmacol.*, 240, 626, 1989.

19. Shank, R.P., Schneider, C.R., and Tighe, J.J., Ion dependence of neurotransmitter uptake: inhibitory effects of ion substrates, *J. Neurochem.*, 49, 381, 1978.

20. Amejdki-Chab, N., Costentin, J., and Bonnet, J.J., Kinetic analysis of the chloride dependence of the neuronal uptake of dopamine and effect of anions on the ability of substrates to compete with the binding of the dopamine uptake inhibitor GBR 12783, *J. Neurochem.*, 58, 793, 1992.

21. Krueger, B.K., Kinetics and block of dopamine uptake in synaptosomes from rat caudate nucleus, *J. Neurochem.*, 55, 260, 1990.

22. McElvain, J.S. and Schenk, J.O., A multisubstrate mechanism of striatal dopamine uptake and its inhibition by cocaine, *Biochem. Pharmacol.*, 43, 2189, 1992.

23. Rostene, W., Boja, J.W., Scherman, D., Carroll, F.I., and Kuhar, M.J., Dopamine transport: pharmacological distinction between the synaptic membrane and vesicular transporter in rat striatum, *Eur. J. Pharmacol.*, 281, 175, 1992.

24. Kitayama, S., Shimada, S., Xu, H., Markham, L., Donovan, D.M., and Uhl, G.R., Dopamine transporter site-directed mutations differentially alter substrate transport and cocaine binding, *Proc. Natl. Acad. Sci. U.S.A.*, 89, 7782, 1992.

25. Lee, F.J.S., Pristupa, Z.B., Ciliax, B.J., Levey, A.L., and Niznik, H.B., The dopamine transporter carboxyl-terminal tail. Truncation/substitution mutants selectively confer high affinity dopamine uptake while attenuating recognition of the ligand binding domain, *J. Biol. Chem.*, 271, 20885, 1996.

26. Volz, T.J. and Schenk, J.O., A comprehensive atlas of the topography of functional groups of the dopamine transporter, *Synapse*, 58, 72, 2005.

27. Heikkila, R.E., Cabbat, F.S., and Duviosin, R.C., Motor activity and rotational behavior after analogs of cocaine: correlation with dopamine uptake blockade, *Commun. Psychopharm.*, 3, 285, 1979.

28. Heikkila, R.E., Manzino, L., and Cabbat, F.S., Stereospecific effects of cocaine derivatives on ^{3}H-dopamine uptake: correlations with behavioral effects, *Subst. Use Misuse*, 2, 115, 1981.

29. Reith, M.E.A., Sershen, H., and Lajtha, A., Saturable [^{3}H]cocaine binding in central nervous system of the mouse, *Life Sci.*, 27, 1055, 1980.

30. Kennedy, L.T. and Hanbauer, I., Sodium-sensitive cocaine binding to rat striatal membrane: possible relationship to dopamine uptake sites, *J. Neurochem.*, 41, 172, 1983.

31. Ritz, M.C., Lamb, R.J., Goldberg, S.R., and Kuhar, M.J., Cocaine receptors on dopamine transporters are related to self-administration of cocaine, *Science*, 237, 1219, 1987.

32. Bergman, J., Madras, B.K., Johnson, S.E., and Spealman, R.D., Effects of cocaine and related drugs in nonhuman primates. III. Self-administration by squirrel monkeys, *J. Pharmacol. Exp. Ther.*, 251, 150, 1989.

33. Scheffel, U., Boja, J.W., and Kuhar, M.J., Cocaine receptors: *in vivo* labeling with ^{3}H-(-) cocaine, ^{3}H-WIN 35,065-2 and ^{3}H-35,428, *Synapse*, 4, 390, 1989.

34. Fowler, J.S., Volkow, N.D., Wolf, A.P., Dewey, S.L., Schlyer, D.J., MacGregor, R.R., Hitzmann, R., Logan, J., Bendriem, B., Gatley, S.J., and Christman, D.R., Mapping cocaine binding sites in human and baboon brain *in vivo*, *Synapse*, 4, 371, 1989.

35. Volkow, N.D., Fowler, J.S., Wolf, A.P., Wang, G.J., Logan, J., MacGregor, D.J., Dewey, S.L., Schlyer, D.J., and Hitzmann, R., Distribution and kinetics of carbon-11-cocaine in the human body measured with PET, *J. Nucl. Med.*, 33, 521, 1992.

36. Scheffel, U., Pogun, S., Stathis, A., Boja, J.W., and Kuhar, M.J., *J. Pharmacol. Exp. Ther.*, 257, 954, 1992.

37. Cline, E.J., Scheffel, U., Boja, J.W., Mitchell, W.M., Carroll, F.I., Abraham, P., Lewin, A.H., and Kuhar, M.J., *In vivo* binding of [^{125}I]RTI-55 to dopamine and serotonin transporters in rat brain, *Synapse*, 12, 37, 1992.

38. Scheffel, U., Dannals, R.F., Cline, E.J., Ricaurte, G.A., Carroll, F.I., Abraham, P., Lewin, A.H., and Kuhar, M.J., [$^{123/125}$I]RTI-55, an *in vivo* label for the serotonin transporter, *Synapse*, 11, 134, 1992.

39. Carroll, F.I., Rahman, M.A., Abraham, P., Parham, K., Lewin, A.H., Dannals, R.F., Shaya, E., Scheffel, U., Wong, D.F., Boja., J.W., and Kuhar, M.J., [^{123}I]3-4(-iodophenyl)tropan-2-caroxylic acid methyl ester (RTI-55), a unique cocaine receptor ligand for imaging the dopamine and serotonin transporters *in vivo*, *Med. Chem. Res.*, 1, 289, 1991.

40. Neumeyer, J.L., Wang, S., Milius, R.M., Baldwin, R.M., Zea-Ponce, Y., Hoffer, P.B., Sybirska, E., Al-tikriti, M., Charney, D.S., Malison, R.T., Laruelle, M., and Innis, R.B., [^{123}I]-2-carbomethoxy-3-(-4-iodophenyl)tropane: high affinity SPECT radiotracer of monoamine reuptake sites in brain, *J. Med. Chem.*, 34, 3144, 1991.

41. Innis, R., Baldwin, R.M., Sybirska, E., Zea, Y., Laruelle, M., Al-Tikriti, M., Charney, D., Zoghbi, S., Wisniewski, G., Hoffer, P., Wang, S., Millius, R., and Neumeyer, J., Single photon emission computed tomography imaging of monoamine uptake sites in primate brain with [^{123}I]CIT, *Eur. J Pharmacol.*, 200, 369, 1991.

42. Shaya, E.K., Scheffel, U., Dannals, R.F., Ricaurte, G.A., Carroll, F.I., Wagner, H.N., Jr., Kuhar, M.J., and Wong, D.F., *In vivo* imaging of dopamine reuptake sites in the primate brain using single photon emission computed tomography (SPECT) and iodine-123 labeled RTI-55, *Synapse*, 10, 169, 1992.

43. Boja, J.W., Mitchell, W.M., Patel, A., Kopajtic, T.A., Carroll, F.I., Lewin, A.H., Abraham, P., and Kuhar, M.J., High affinity binding of [^{125}I]RTI-55 to dopamine and serotonin transporters in rat brain, *Synapse*, 12, 27, 1992.

44. Boja, J.W., Cadet, J.L., Kopajtic, T.A., Lever, J., Seltzman, H.H., Wyrick, C.D., Lewin, A.H., Abraham, P., and Carroll, F.I., Selective labeling of the dopamine transporter by the high affinity ligand 3-(4-[^{125}I]iodophenyltropane-2-carboxylic acid isopropyl ester, *Mol. Pharmacol.*, 47, 779, 1995.

45. Staley, J.K., Boja, J.W., Carroll, F.I., Seltzman, H.H., Wyrick, C.D., Lewin, A.H., Abraham, P., and Mash, D.C., Mapping dopamine transporters in the human brain with novel selective cocaine analog [^{125}I]RTI-121, *Synapse,* 21, 364, 1995.

46. Ciliax, B.J., Heilman, C., Demchyshyn, L.L., Pristupa, Z.B., Ince, E., Hersch, S.M., Niznik, H.B., and Levey, A.I., The dopamine transporter: immunochemical characterization and localization in brain, *J. Neurosci.,* 15, 1714, 1995.

47. Javitch, J.A., Blaustein, R.O., and Snyder, S.H., [^{3}H]Mazindol binding associated with neuronal dopamine and norepinphrine uptake sites, *Mol. Pharmacol.*, 26, 35, 1984.

48. Dubocovich, M.L. and Zahniser, N.R., Binding characteristics of dopamine uptake inhibitor [^{3}H]nomifensine to striatal membranes, *Biochem. Pharmacol.*, 34, 1137, 1985.

49. Anderson, P.H., Biochemical and pharmacological characterization of [^{3}H]GBR 12935 binding *in vitro* to rat striatal membranes: labeling of the dopamine uptake complex, *J. Neurochem.*, 48, 1887, 1987.

50. Schoemaker, H., Pimoule, C., Arbilla, S., Scatton, B., Javoy-Agid, F., and Langer, S.Z., Sodium dependent [^{3}H]cocaine binding associated with dopamine uptake sites in the rat striatum and human putamen decrease after dopamine denervation and in Parkinson's disease, *Naunyn-Schmiedeberg's Arch. Pharmacol.*, 329, 227, 1985.

51. Calligaro, D.O. and Eldefraei, M.E., High affinity stereospecific binding of [^{3}H]cocaine in striatum and its relationship to the dopamine transporter, *Membr. Biochem.*, 7, 87, 1988.

52. Madras, B.K., Fahey, M.A., Bergman, J., Canfield, D.R., and Spealman, R.D., Effects of cocaine and related drugs in nonhuman primates. I. [^{3}H]Cocaine binding sites in caudate-putamen, *J. Pharmacol. Exp. Ther.*, 251, 131, 1989.

53. Ritz, M.C., Boja, J.W., Zaczek, R., Carroll, F.I., and Kuhar, M.J., ^{3}H WIN 35,065-2: a ligand for cocaine receptors in striatum, *J. Neurochem.*, 55, 1556, 1990.

54. Madras, B.K., Spealman, R.D., Fahey, M.A., Neumeyer, J.L., Saha, J.K., and Milius, R.A., Cocaine receptors labeled by [^{3}H]2-carbomethoxy-3-(4-fluorophenyl)tropane, *Mol. Pharmacol.*, 36, 518, 1989.

55. Boja, J.W., Markham, L., Patel, A., Uhl, G., and Kuhar, M.J., Expression of a single dopamine transporter cDNA can confer two cocaine binding sites, *Neuroreport*, 3, 247, 1992.

56. Madras, B.K. and Kaufman, M.J., Cocaine accumulates in dopamine-rich regions of primate brain after I.V. administration: comparison with mazindol distribution, *Synapse,* 18, 261, 1994.

57. Pögün, S., Scheffel, U., and Kuhar, M.J., Cocaine displaces [^{3}H]WIN 35,428 binding to dopamine uptake sites *in vivo* more rapidly than mazindol or GBR 12909, *Eur. J. Pharmacol.*, 198, 203, 1991.

58. Stathis, M., Scheffel, U., Lever, S.Z., Boja, J.W., Carroll, F.I., and Kuhar, M.J., Rate of binding of various inhibitors at the dopamine transporter *in vivo, Psychopharmacology,* 119, 376, 1995.

59. Giros, B., Jaber, M., Jones, S.R., Wightman, R.M., and Caron, M.G., Hyperlocomotion and indifference to cocaine and amphetamine in mice lacking the dopamine transporter, *Nature,* 379, 606, 1996.

60. Sora, I., Hall, F.S., Andrews, A.M., Itokawa, M., Li, X., Wei, H., Wichems, C., Lesch, K., Murphy, D.L., and Uhl, G.R., Molecular mechanisms of cocaine reward: combined dopamine and serotonin transporter knockouts eliminate cocaine place preference, *PNAS*, 98, 5300, 2001.

61. Carboni, E., Spielewoy, C., Vacca, C., Norten-Bertrand, M., Giros, B., and DiChiara, G., Cocaine and amphetamine increase extracellular dopamine in the nucleus accumbens of mice lacking the dopamine transporter gene, *J. Neurosci.,* 21, 1, 2001.

62. Mead, A.N., Rocha, B.A., Donovan, D.M., and Katz, J.L., Intravenous cocaine induced-activity and behavioral sensitization in norepinephrine-, but not dopamine-transporter knockout mice, *Eur. J. Neurosci.*, 16, 514, 2002.

63. Hall, F.S., Sora, I., Drgonova, J., Li, X.F., Goeb, M., and Uhl, G.R., Molecular mechanisms underlying the rewarding effects of cocaine, *Ann. N.Y. Acad. Sci.*, 1025, 47, 2004.

64. Ungerstedt, U., Striatal dopamine release after amphetamine or nerve degeneration revealed by rotational behavior, *Acta Physiol. Scand.*, 367, 49, 1971.

65. Masuoka, D.T., Alcaraz, A.F., and Schott, H.F., [^{3}H]Dopamine release by d-amphetamine from striatal synaptosomes of reserpinized rats, *Biochem. Pharmacol.*, 31, 1969, 1982.

66. Kuczenski, R., Biochemical actions of amphetamine and other stimulants, in *Stimulants: Neurochemical, Behavioral, and Clinical Perspectives*, Creese, I., Ed., Raven Press, New York, 1983, 31.

67. Bowyer, J.F., Spuler, K.P., and Weiner, N., Effects of phencyclidine, amphetamine and related compounds on dopamine release from and uptake into striatal synaptosomes, *J. Pharmacol. Exp. Ther.*, 229, 671, 1984.

68. Moghaddam, B., Roth, R.H., and Bunny, B.S., Characterization of dopamine release in the rat medial prefrontal cortex as assessed by *in vivo* microdialysis: comparison to the striatum, *Neuroscience,* 36, 669, 1990.

69. Robertson, G.S., Damsma, G., and Fibiger, H.C., Characterization of dopamine release in the substantia nigra by *in vivo* microdialysis in freely moving rats, *J. Neurosci.,* 11, 2209, 1991.

70. Schmidt, C.J. and Gibb, J.W., Role of the dopamine uptake carrier in the neurochemical response to methamphetamine: effects of amfonelic acid, *Eur. J. Pharmacol.*, 109, 73, 1985.

71. Johnson, M.P., Hoffman, A.J., and Nichols, D.E., Effects of the enantimers of MDA, MDMA, and related analogues of [^{3}H]serotonin and [^{3}H]dopamine release from superfused rat brain slices, *Eur. J. Pharmacol.*, 132, 269, 1986.

72. Steele, T.D., Nichols, D.E., and Yim, G.K., Stereochemical effects of 3,4-methylendioxymethamphetamine (MDMA) and related amphetamine derivatives on inhibition of uptake of [^{3}H]monoamines into synaptosomes from different regions of rat brain, *Biochem. Pharmacol.*, 36, 2297, 1987.

73. Yamamoto, B.K. and Spanos, L.J., The acute effects of methylenedioxymethamphetamine on dopamine release in awake-behaving rat, *Eur. J. Pharmacol.*, 148, 195, 1988.

74. Nash, J.F. and Nichols, D.E., Microdialysis studies on 3,4-methylenedioxyamphetamine and structurally related analogues, *Eur. J. Pharmacol.*, 200, 53, 1991.

75. Johnson, M.P., Conarty, P.F., and Nichols, D.E., [^{3}H]Monoamine releasing and uptake inhibition properties of 3,4-methylenedioxymethamphetamine and *p*-chloroamphetamine analogues, *Eur. J. Pharmacol.*, 200, 9, 1991.

76. Azzaro, A.J., Ziance, R.J., and Rutledge, C.O., The importance of neuronal uptake of amines for amphetamine-induced release of ^{3}H-norepinephrine from isolated brain tissue, *J. Pharmacol. Exp. Ther.,* 189, 110, 1974.

77. Hadfield, M.G., A comparison of *in vivo* and *in vitro* amphetamine on synaptosomal uptake of dopamine in mouse striatum, *Res. Commun. Chem. Mol. Pathol. Pharmacol.,* 48, 183, 1985.

78. Arnold, E.B., Molinoff, P.B., and Rutledge, C.O., The release of endogenous norepinephrine and dopamine from cerebral cortex by amphetamine, *J. Pharmacol. Exp. Ther.*, 202, 544, 1977.

79. Fisher, J.F. and Cho, A.K., Chemical release of dopamine from striatal homogenates: evidence for an exchange diffusion model, *J. Pharmacol. Exp. Ther.,* 208, 203, 1979.

80. Liang, N.Y. and Rutledge, C.O., Comparison of the release of [^{3}H]dopamine from isolated corpus striatum by amphetamine, fenfluramine and unlabeled dopamine, *Biochem. Pharmacol.*, 31, 983, 1982.

81. Zaczek, R., Culp, S., and De Souza, E.B., Intrasynaptosomal sequestration of [^{3}H]amphetamine and [^{3}H]methylenedioxyamphetamine: characterization suggests the presence of a factor responsible for maintaining sequestration, *J. Neurochem.*, 54, 195, 1990.

82. Zaczek, R., Culp, S., and De Souza, E.B., Interactions of [^{3}H]amphetamine with rat brain synaptosomes. II. Active, *J. Pharmacol. Exp. Ther.*, 257, 830, 1991.

83. Jacocks, H.M. and Cox, B.K., Serotonin-stimulated [^{3}H]dopamine via reversal of the dopamine transporter in rat striatum and nucleus accumbens: a comparison with release elicited by potassium, N-methyl-D-aspartic acid, glutamic acid and D-amphetamine, *J. Pharmacol. Exp. Ther.*, 262, 356, 1992.

84. Sulzer, D., Maidment, N.T., and Rayport, S., Amphetamine and other weak bases act to promote reverse transport of dopamine in ventral midbrain neurons, *J. Neurochem.*, 60, 527, 1993.

85. Eshleman, A.J., Henningsen, R.A., Neve, K.A., and Janowsky, A., Release of dopamine via the human transporter, *Mol. Pharmacol.*, 45, 312, 1994.

86. Kahlig, K.M., Binda, F., Khoshbouei, H., Blakely, R.D., McMahon, D.G., Javitch, J.A., and Galli, A., Amphetamine induces dopamine efflux through a dopamine transporter channel, *PNAS,* 102, 3495, 2005.

87. Pifl, C., Drobny, H., Reither, H., Hornykiewicz, O., and Singer, E.A., Mechanism of the dopamine-releasing actions of amphetamine and cocaine: plasmalemmal dopamine transporter versus vesicular monoamine transporter, *Mol. Pharmacol.*, 47, 368, 1995.

88. Di Chiara, G. and Imperato, A., Opposite effects of mu and kappa opiate agonists on dopamine release in the nucleus accumbens and in the dorsal caudate of freely moving rat, *J. Pharmacol. Exp. Ther.*, 244, 1067, 1988b.

89. Di Chiara, G. and Imperato, A., Opposite effects of mu and kappa opiate agonists on dopamine release in the nucleus accumbens and in the dorsal caudate of freely moving rat, *J. Pharmacol. Exp. Ther.*, 244, 1067, 1988b.

90. Hurd, Y.L., Weiss, F., Koob, G., and Ungerstedt, U., Cocaine reinforcement and extracellular dopamine overflow in the rat nucleus accumbens: an *in vivo* microdialysis study, *Brain Res.*, 498, 199, 1989.

91. Kalivas, P.W., Winderlov, E., Stanley, D., Breese, G.R., and Prange, A.J., Jr., Enkephalin action on the mesolimbicdopamine system: a dopamine-dependent and dopamine-independent increase in locomotor activity, *J. Pharmacol. Exp. Ther.*, 227, 229, 1983.

92. Pettit, H.O., Ettenberg, A., Bloom, F.E., and Koob, G.F., Destruction of dopamine in the nucleus accumbens selectively attenuates cocaine but not heroin self-administration in rats, *Psychopharmacology*, 84, 167, 1984.

93. Sellings, L.H. and Clarke, P.B., Segregation of amphetamine reward and locomotor stimulation between nucleus accumbens medial shell and core, *J. Neurosci.*, 23, 6295, 2003.

94. Di Chiara, G. and North, A.R., Neurobiology of opiate abuse, *TIPS*, 13, 185, 1992.

95. Koob, G.F., Drugs of abuse: anatomy, pharmacology and function of reward pathways, *TIPS*, 13, 177, 1992.

96. Spyraki, C., Fibiger, H.C., and Phillips, A.G., Attenuation of heroin reward in rats by disruption of the mesolimbic dopamine system, *Psychopharmacology*, 79, 278, 1983.

97. Zito, K.A., Vickers, G., and Roberts, D.C.S., Disruption of cocaine and heroin self-administration following kianic acid lesions of the nucleus accumbens, *Pharmacol. Biochem. Behav.*, 23, 1029, 1985.

98. Bozarth, M.A. and Wise, R.A., Heroin reward is dependent on a dopaminergic substrate, *Life Sci.*, 29, 1881, 1981.

99. Kornetsky, C. and Porrino, L.J., Brain mechanisms of drug-induced reinforcement, in *Addictive States*, O'Brien, C.P. and Jaffe, J.H., Eds., Raven Press, New York, 1992, 59.

100. Reith, M.E.A., Meisler, B.E., Sershen, H., and Lajtha, A., Structural requirements for cocaine congeners to interact with dopamine and serotonin uptake sites in mouse brain and to induce stereotyped behavior, *Biochem. Pharmacol.*, 35, 1123, 1986.

101. Gysling, K. and Wang, R.Y., Morphine-induced activation of A10 dopamine neurons in the rat brain, *Brain Res.*, 277, 119, 1983.

102. Matthews, R.T. and German, D.C., Electrophysiological evidence for excitation of rat ventral tegmental area dopamine neurons by morphine, *Neuroscience*, 11, 617, 1984.

103. Das, D., Rogers, J., and Michael-Titus, A.T., Comparative study of the effects of mu, delta and kappa opioid agonists on 3H-dopamine uptake in the rat striatum and nucleus accumbens, *Neuropharmacology*, 33, 221, 1994.

104. Schenk, S., Partridge, B., and Shippenberg, T.S., Reinstatement of extinguished drug-taking behavior in rats: effect of the kappa-opioid receptor agonist, U69593, *Psychopharmacology*, 151, 85, 2000.

105. Izenwasser, S., Newman A.H., Cox, B.M., and Katz, J.L., The cocaine-like behavioral effects of meperidine are mediated by activity at the dopamine transporter, *Eur. J. Pharmacol.*, 297, 9, 1996.

106. Bergstrom, K.A., Jolkkonen, J., Kuikka, J.T., Akerman, K.K., Viinamaki, H., Airaksinen, O., Lansimies, E., and Tiihonen, J., Fentanyl decreases beta-CIT binding to the dopamine transporter, *Synapse*, 29, 413, 1998.

107. Simantov, R., Chronic morphine alters dopamine transporter density in the rat brain: possible role in the mechanism of drug addiction, *Neurosci. Lett.*, 163, 121, 1993.

108. Spielwoy, C., Gonon, F., Roubert, C., Fauchey, V., Jaber, M., Caron, M.G., Roques, B.P., Hamon, M., Betancur, C., Maldonado, R., and Giros, B., Increased rewarding properties of morphine in dopamine-transporter knockout mice, *Eur. J. Neurosci.*, 12, 1827, 2000.

109. Svingos, A.L., Clarke, C.L., and Pickel, V.M., Localization of delta-opioid receptor and dopamine transporter in the nucleus accumbens shell: implications for opiate and psychostimulant cross sensitization, *Synapse*, 34, 1, 1999.

110. Izenwasser, S. and Cox, B.M., Daily cocaine treatment produces a persistent reduction of [3H]dopamine uptake *in vitro* in the rat nucleus accumbens but not the striatum, *Brain Res.*, 531, 338, 1990.

111. Gudehithlu, K.P. and Bhargava, H.N., Modification of characteristics of dopamine transporter in brain regions and spinal cord of morphine tolerant and abstinent rats, *Neuropharmacology*, 35, 169, 1996.

112. Spangler, R., Goddard, N.L., Avena, N.M., Hoebel, B.G., and Leibowitz, S.F., Elevated D3 dopamine receptor mRNA in dopaminergic and dopaminoceptive regions of the rat brain in response to morphine, *Brain Res. Mol. Brain Res.*, 111, 74, 2003.

113. Kish, S.J., Kalasinsky, K.S., Derach, P., Schmunk, G.A., Guttman, M., Ang, L., Adams, V., Furukawa, Y., and Haycock, J.W., Striatal dopaminergic and serotonergic markers in human heroin users, *Neuropsychopharmacology*, 24, 561, 2001.

114. Carboni, E., Imperato, A., Perezzani, L., and Di Chiara, G., Amphetamine, cocaine, phencyclidine and nomifensine increase extracellular dopamine concentrations preferentially in the nucleus accumbens of freely moving rats, *Neuroscience*, 28, 653, 1989.

115. Hernandez, L., Auerbach, S., and Hoebel, B.G., Phencyclidine (PCP) injected into the nucleus accumbens increases extracellular dopamine and serotonin as measured by microdialysis, *Life Sci.*, 42, 1713, 1988.

116. Hondo, H., Yonezawa, Y., Nakahara, T., Nakamura, K., Hirano, M., Uchimura, H., and Tashiro, N., Effect of Phencyclidine on dopamine release in the rat prefrontal cortex; an *in vivo* microdialysis study, *Brain Res.*, 633, 337, 1994.

117. Hata, N., Nishikawa, T., Umino, A., and Takahashi, K., Evidence for involvement of N-methyl-D-aspartate receptor in tonic inhibitory control of dopaminergic transmission in the rat medial frontal cortex, *Neurosci. Lett.*, 120, 101, 1990.

118. Freeman, A.S. and Bunney, B.S., The effects of phencyclidine and N-allylnormetazocine on mid-brain dopamine neuronal activity, *Eur. J. Pharmacol.*, 104, 287, 1984.

119. Gariano, R.F. and Groves, P.M., Burst firing induced in mid-brain dopamine neurons by stimulation of the medial prefrontal and anterior cingulate cortices, *Brain Res.*, 462, 194, 1988.

120. Suaad-Chagny, M.F., Chergui, K., Chouvet, G., and Gonon, F., Relationship between dopamine release in the rat nucleus accumbens and the discharge activity of dopaminergic neurons during local *in vivo* application of amino acids in the ventral tegmental area, *Neuroscience*, 49, 63, 1992.

121. Fagg, G.E., Phencyclidine and related drugs bind to the activated N-methyl-D-aspartate receptor-channels complex in rat membranes, *Neurosci. Lett.*, 76, 221, 1987.

122. Mount, H., Boksa, P., Chadieu, I., and Quirion, R., Phencyclidine and related compounds evoked [3H] dopamine release from rat mesencephalon cell cultures by mechanisms independent of the phencyclidine receptor, sigma binding site, or dopamine uptake site, *Can. J. Physiol. Pharmacol.*, 68, 1200, 1990.

123. Bowyer, J.F., Spuhler, K.P., and Weiner, N., Effects of phencyclidine, amphetamine, and related compounds on dopamine release from and uptake into striatal synaptosomes, *J. Pharmacol. Exp. Ther.*, 229, 671, 1984.

124. Baba, A., Yamamoto, T., Yamamoto, H., Suzuki, T., and Moroji, T., Effects of the major metabolite of phencyclidine, the trans isomer of 4-phenyl-4-(1-piperidinyl) cyclohexanol, on [3H]N-(1-[2-thienyl]cyclohexyl)-3,4-piperidine([3H}TPC) binding and [3H] dopamine uptake in the rat brain, *Neurosci. Lett.*, 182, 119, 1994.

125. Garey, R.E. and Heath, R.G., The effects of phencyclidine on the uptake of 3H-catecholamines by rat striatal and hypothalamic synaptosomes, *Life Sci.*, 18, 1105, 1976.

126. Smith, R.C., Meltzer, H.Y., Arora, R.C., and Davis, J.M., Effects of phencyclidine on H-catecholamines and H-serotonin uptake in synaptosomal preparations from the rat brain, *Biochem. Pharmacol.*, 26, 1435, 1977.

127. Gerhardt, G.A., Pang, K., and Rose, G.M., *In vivo* electrochemical demonstration of presynaptic actions of phencyclidine in rat caudate nucleus, *J. Pharmacol. Exp. Ther.*, 241, 714, 1987.

128. Vignon, J., Pinet, V., Cerruti, C., Kamenka, J., and Chicheportiche, R., [3H]N-1(2-Benzo(b)thiophenyl)cyclohexyl]piperidinme ([3H]BTCP): a new phencyclidine analog selective for the dopamine uptake complex, *Eur. J. Pharmacol.*, 148, 427, 1988.

129. Kuhar, M.J., Boja, J.W., and Cone, E.J., Phencyclidine binding to striatal cocaine receptors, *Neuropharmacology*, 29, 295, 1990.

130. Cook, C.E., Perez, R.M., Jeffcoat, A.R., and Brine, D.R., Phencyclidine disposition in humans after small doses of radiolabeled drug, *Fed. Proc.*, 42, 2566, 1983.

131. Schiffer, W.K., Logan, J., and Dewey, S.L., Positron emission tomography studies of potential mechanisms underlying phencyclidine-induced alterations in striatal dopamine, *Neuropharmacology*, 28, 2192, 2003.

132. Carlezon, W.A., Jr. and Wise, R.A., Rewarding actions of phencyclidine and related drugs in the nucleus accumbens shell and frontal cortex, *J. Neurosci.*, 16, 3112, 1996.

133. Carlezon, W.A., Jr. and Wise, R.A., Habit-forming actions of nomifensine in the nucleus accumbens, *Psychopharmacology*, 122, 194, 1995.

134. Steinpreis, R.E. and Salamone, J.D., The role of nucleus accumbens dopamine in the neurochemical and behavioral effects of phencyclidine: a microdialysis and behavioral study, *Brain Res.*, 612, 263, 1993.

135. Hanania, T. and Zahniser, N.R., Locomotor activity induced by noncompetitive NMDA receptor antagonists versus dopamine transporter inhibitors: opposite strain differences in inbred long-sleep and short-sleep mice, *Alcohol Clin. Exp. Res.*, 26, 431, 2002.

136. Nicoll, R.A. and Alger, B.E., The brain's own marijuana, *Sci. Am.*, 291, 68, 2005.

137. Chaperon, F. and Thiebot, M.H., Behavioral effects of cannabinoid agents in animals, *Crit. Rev. Neurobiol.*, 13, 243, 1999.

138. Ameri, A., The effects of cannabinoids on the brain, *Prog. Neurobiol.*, 59, 315, 1999.

139. Chen, J., Paredes, W., Lowinson, J.H., and Gardner, E.L., Δ^9-Tetrahydrocannabinol enhances presynaptic dopamine efflux in the medial prefrontal cortex, *Eur. J. Pharmacol.*, 190, 259, 1990.

140. Pistis, M., Ferraro, L., Flore, G., Tanganelli, S., Gessa, G.L., and Devoto, P., Delta(9)-tetrahydrocannabinol decreases extracellular GABA and increases extracellular glutamate and dopamine levels in the prefrontal cortex: an *in vivo* microdialysis study, *Brain Res.*, 948, 155, 2002.

141. Ng Cheong Ton, J.M., Gerhardt, G.A., Friedmann, M., Etgen, A.M., Rose, G.M., Sharpless, N.S., and Gardner, E.L., Effects of Δ^9-tetrahydrocannabinol on potassium-evoked release of dopamine in the rat caudate nucleus: an *in vivo* electrochemical and *in vivo* microdialysis study, *Brain Res.*, 451, 59, 1988.

142. Chen, J., Paredes, W., Li, J., Smith, D., Lowinson, J., and Gardner, E.L., *In vivo* brain microdialysis studies of Δ^9-tetrahydrocannabinol on presynaptic dopamine efflux in nucleus accumbens of the Lewis rat, *Psychopharmacology*, 102, 156, 1990.

143. Chen, J., Paredes, W., Lowinson, J.H., and Gardner, E.L., Strain-specific facilitation of dopamine efflux by Δ^9-tetrahydrocannabinol in the nucleus accumbens of a rat: an *in vivo* microdialysis study, *Neurosci. Lett.*, 129, 136, 1991.

144. Chen, J., Marmur, R., Pulles, A., Paredes, W., and Gardner, E.L., Ventral tegmental microinjection of Δ^9-tetrahydrocannabinol enhances ventral tegmental somatodendritic dopamine levels but not forebrain dopamine levels: evidence for local neural action by marijuana's psychoactive ingredient, *Brain Res.*, 621, 65, 1993.

145. French, E.D., Dillion, K., and Wu, X., Cannabinoids excite dopamine neurons in the ventral tegmentum and substantia nigra, *Neuroreport*, 8, 649, 1997.

146. Riegel, A.C. and Lupica, C.R., Independent presynaptic and postsynaptic mechanisms regulate endocannabinoid signaling at multiple synapses in the ventral tegmental area, *J. Neurosci.*, 24, 11070, 2004.

147. Melis, M., Perra, S., Muntoni, A.L., Pillolla, G., Lutz, B., Marsicano, G., Di Marzo, V., Gessa, G.L., and Pistis, M., Prefrontal cortex stimulation induces 2-arachidonoyl-glycerol-mediated suppression of excitation in dopamine neurons, *J. Neurosci.*, 24, 10707, 2004.

148. Valjent, E., Pages, C., Rogard, M., Besson, M.J., Maldonado, R., and Caboche, J., Delta 9-tetrahydrocannabinol-induced MAPK/ERK and Elk-1 activation *in vivo* depends on dopaminergic transmission, *Eur. J. Neurosci.*, 14, 342, 2001.

149. Bloom, A.S., Effect of delta-9-tetrahydrocannabinol on the synthesis of dopamine and norepinephrine in mouse brain synaptosomes, *J. Pharmacol. Exp. Ther.*, 221, 97, 1982.

150. Poddar, M.K. and Dewey, W.L., Effects of cannabinoids on catecholamine uptake and release in hypothalamic and striatal synaptosomes, *J. Pharmacol. Exp. Ther.*, 214, 63, 1980.

151. Hershkowitz, M. and Szechtman, H., Pretreatment with Δ^1 tetrahydrocannabinol and psychoactive drugs: effects on uptake of biogenic amines and on behavior, *Eur. J. Pharm.*, 59, 267, 1979.

152. Banerjee, S.P., Snyder, S.H., and Mechoulam, R., Cannabinoids: influence on neurotransmitter uptake in rat brain synaptosomes, *J. Pharmacol. Exp. Ther.*, 194, 74, 1975.

153. Sakurai-Yamashita, Y., Kataoka, Y., Fujiwara, M., Mine, K., and Ueki, S., Delta 9-tetrahydrocannabinol facilitates striatal dopaminergic transmission, *Pharmacol. Biochem. Behav.*, 33, 397, 1989.

154. Dean, B., Bradbury, R., and Copolov, D.L., Cannabis-sensitive dopaminergic markers in postmortem central nervous system: changes in schizophrenia, *Biol. Psychiatry*, 53, 585, 2003.

155. Wu, X. and French, E.D., Effects of chronic delta9-tetrahydrocannabinol on rat midbrain dopamine neurons: an electrophysiological assessment, *Neuropharmacology*, 39, 391, 2000.

156. Gardner, E.L., Paredes, W., Smith, D., Donner, A., Milling, C., Cohen, D., and Morrison, D., Facilitation of brain stimulation reward by Δ^9-tetrahydrocannabinol, *Psychopharmacology*, 96, 142, 1988.

157. Braida, D., Losue, S., Pegorini, S., and Sala, M., Delta9-tetrahydrocannabinol-induced conditioned place preference and intracerebroventricular self-administration in rats, *Eur. J. Pharmacol.*, 506, 63, 2004.

158. Justinova, Z., Tanda, G., Redhi, G.H., and Goldberg, S.R., Self-administration of delta(9)-tetrahydro-cannabinol (THC) by drug naïve squirrel monkeys, *Psychopharmacology,* 169, 135, 2003.

159. Fattore, L., Spano, S., Gregorio, C., Deiana, S., Fadda, P., and Fratta, W., Cannabinoid CB1 antagonist SR 141716A attenuates reinstatement of heroin self-administration in heroin-abstinent rats, *Neuropharmacology,* 48, 1097, 2005.

160. Samuel, D., Lynch, M.A., and Littleton, J.M., Picrotoxin inhibits the effect of ethanol on the spontaneous efflux of [^{3}H]-dopamine from superfused slices of rat corpus striatum, *Neuropharmacology,* 22, 1412, 1983.

161. Shier, W.T., Koda, L.Y., and Bloom, F.E., Metabolism of [^{3}H]dopamine following intracerebroventricular injection in rats pretreated with ethanol or choral hydrate, *Neuropharmacology,* 22, 279, 1983.

162. Russell, V.A., Lamm, M.C., and Taljaard, J.J., Effects of ethanol on [^{3}H]dopamine release in rat nucleus accumbens and striatal slices, *Neurochem. Res.,* 13, 487, 1988.

163. Strombom, U.H. and Liedman, B., Role of dopaminergic neurotransmission in locomotor stimulation by dexamphetamine and ethanol, *Psychopharmacology,* 78, 271, 1982.

164. Murphy, J.M., McBride, W.J., Lumeng, L., and Li, T.K., Monoamine and metabolite levels in CNS regions of the P line of alcohol-preferring rats after acute and chronic ethanol treatment, *Pharmacol. Biochem. Behav.,* 19, 849, 1983.

165. Di Chiara, G. and Imperato, A., Ethanol preferentially stimulates dopamine release in the nucleus accumbens of freely moving rats, *Eur. J. Pharmacol.,* 115, 131, 1985.

166. Imperato, I. and Di Chiara, G., Preferential stimulation of dopamine release in the nucleus accumbens of freely moving rats by ethanol, *J. Pharmacol. Exp. Ther.,* 239, 219, 1986.

167. Yoshimoto, K., McBride, W.J., Lumberg, L., and Li, T.K., Ethanol enhances the release of dopamine and serotonin in the nucleus accumbens, *Alcohol,* 9, 17, 1992.

168. McBride, W.J., Murphy, J.M., Gatto, G.J., Levy, A.D., Yoshimoto, K., Lumeng, L., and Li, T.K., CNS mechanisms of alcohol self-administration, *Alcohol Alcohol.,* Suppl. 2, 463, 1993.

169. Samson, H.H. and Hodge, C.W., The role of the mesoaccumbens dopamine system in ethanol reinforcement: studies using the techniques of microinjection and voltammetry, *Alcohol Alcohol.,* Suppl. 2, 469, 1993.

170. Weiss, F., Lorang, M.T., Bloom, F.E., and Koob, G.F., Oral alcohol self-administration stimulates dopamine release in the nucleus accumbens: genetic and motivational determinants, *J. Pharmacol. Exp. Ther.,* 267, 250, 1993.

171. Kiianmaa, K., Nurmi, M., Nykanen, I., and Sinclair, J.D., Effect of ethanol on extracellular dopamine in the nucleus accumbens of alcohol-preferring AA and alcohol-avoiding ANA rats, *Pharmacol. Biochem. Behav.,* 52, 29, 1995.

172. Tan, A.Y., Dular, R., and Innes, I.R., Alcohol feeding alters [^{3}H]dopamine uptake into rat cortical and brain stem synaptosomes, *Prog. Biochem. Pharmacol.,* 18, 224, 1981.

173. Mayfield, R.D., Maiya, R., Keller, D., and Zahniser, N.R., Ethanol potentiates the function of the human dopamine transporter expressed in *Xenopus* oocytes, *J. Neurochem.,* 79, 1070, 2001.

174. Maiya, R., Buck, K.J., Harris, R.A., and Mayfield, R.D., Ethanol-sensitive sites on the human dopamine transporter, *J. Biol. Chem.,* 277, 30724, 2002.

175. Robinson, D.L., Volz, T.J., Schenk, J.O., and Wightman, R.M., Acute ethanol decreases dopamine transporter velocity in rat striatum: *in vivo* and *in vitro* electrochemical measurements, *Alcohol Clin. Exp. Res.,* 29, 746, 2005.

176. Lynch, M.A., Samuel, D., and Littleton, J.M., Altered characteristics of [^{3}H]dopamine release from superfused slices of corpus striatum obtained from rats receiving ethanol *in vivo, Neuropharmacology,* 24, 479, 1985.

177. Yan, Q.S., Ethanol-induced nonexocytotic [^{3}H]dopamine release from rat nucleus accumbens slices, *Alcohol,* 27, 127, 2002.

178. Yim, H.J. and Gonzales, R.A., Ethanol-induced increase in dopamine extracellular concentrations in rat nucleus accumbens are accounted for by increased release and not uptake inhibition, *Alcohol,* 22, 107, 2000.

179. Savelieva, K.V., Caudle, W.M., Findlay, G.S., Caron, M.G., and Miller, G.W., Decreased ethanol preferences and consumption in dopamine transporter female knock-out mice, *Alcohol. Clin. Exp. Res.,* 26, 2002.

180. Hearn, W.L., Flynn, D.D., Hime, G.W., Rose, S., Cofino, J.C., Mantero-Atienza, E., Wetli, C.V., and Mash, D.C., Cocaethylene; a unique metabolite displays high affinity for the dopamine transporter, *J. Neurochem.*, 56, 698, 1991.

181. Jatlow, P., Elsworth, J.D., Bradberry, C.W., Winger, G., Taylor, J.R., Russell, R., and Roth, R.H., Cocaethylene: a neuropharmacologically active metabolite associated with concurrent cocaine-ethanol ingestion, *Life Sci.*, 48, 1781, 1991.

182. Woodward, J.J., Mansbach, R., Carroll, F.I., and Balster, R.L., Cocaethylene inhibits dopamine uptake and produces cocaine-like actions in drug discrimination studies, *Eur. J. Pharmacol.*, 197, 235, 1991.

183. Lewin, A.H., Gao, Y., Abraham, P., Boja, J.W., Kuhar, M.J., and Carroll, F.I., The effect of 2-substitution on binding affinity at the cocaine receptor, *J. Med. Chem.*, 35, 135, 1992.

184. Bradberry, C.W., Nobiletti, J.B., Elsworth, J.D., Murphy, B., Jatlow, P., and Roth, R.H., Cocaine and cocaethylene; microdialysis comparison of brain drug levels and effects on dopamine and serotonin, *J. Neurochem.*, 60, 1429, 1993.

185. Iyer, R.N., Nobiletti, J.B., Jatlow, P.I., and Bradberry, C.W., Cocaine and cocaethylene: effects of extracellular dopamine in the primate, *Psychopharmacology,* 120, 150, 1995.

186. Katz, J.I., Terry, P., and Witkin, J.M., Comparative behavioral pharmacology and toxicology of cocaine and its ethanol-derived metabolite, cocaine ethyl-ester (cocaethylene), *Life Sci.*, 50, 1351, 1992.

187. Hearn, W.L., Rose, S.L., Wagner, J., Ciarleglio, A.C., and Mash, D.C., Cocaethylene is more potent than cocaine in mediating lethality, *Pharmacol. Biochem. Behav.*, 39, 531, 1991.

188. Meehan, S.M. and Schechter, M.D., Cocaethylene-induced lethality in mice is potentiated by alcohol, *Alcohol,* 12, 383, 1995.

189. Wilson, L.D., Jeromin, J., Garvey, L., and Dorbandt, A., Cocaine, ethanol, and cocaethylene cardiotoxicity in an animal model of cocaine and ethanol abuse, *Acad. Emerg. Med.,* 8, 211, 2001.

190. Pennings, E.J., Leccese, A.P., and Wolff, F.A., Effects of concurrent use of alcohol and cocaine, *Addiction,* 97, 773, 2002.

191. Hart, C.L., Jatlow, P., Sevarino, K.A., and McCance-Katz, E.F., Comparison of intravenous cocaethylene and cocaine in humans, *Psychopharmacology,* 149, 153, 2001.

192. Knackstedt, L.A., Samimi, M.M., and Ettenberg, A., Evidence for opponent-process actions of intravenous cocaine and cocaethylene, *Pharmacol. Biochem. Behav.*, 72, 931, 2002.

193. Damsma, G., Westernik, B.H., de Vries, J.B., and Horn, A.S., The effect of systemically applied cholinergic drugs on the striatal release of dopamine and its metabolites, as determined by automated microdialysis in conscious rats, *Neurosci. Lett.,* 89, 349, 1988.

194. Westfall, T.C., Effect of nicotine and other drugs on the release of ^{3}H-norepinephrine and ^{3}H-dopamine from rat brain slices, *Neuropharmacology,* 13, 693, 1974.

195. Marien, M., Brien, J., and Jhamandas, K., Regional release of [^{3}H]dopamine from rat brain *in vitro*: effects of opioids on release induced by potassium nicotine, and L-glutamic acid, *Can. J. Physiol. Pharmacol.*, 61, 43, 1983.

196. Rapier, C., Lunt, G.G., and Wonnacott, S., Stereoselective nicotine-induced release of dopamine from striatal synaptosomes: concentration dependence and repetitive stimulation, *J. Neurochem.*, 50, 1123, 1988.

197. Izenwasser, S., Jacocks, H.M., Rosenberger, J.G., and Cox, B.M., Nicotine indirectly inhibits [^{3}H]dopamine uptake at concentrations that do not directly promote [^{3}H]dopamine release in rat striatum, *J. Neurochem.*, 56, 603, 1991.

198. Dwoskin, L.P., Leibee, L.L., Jewell, A.L., Fang, Z., and Crooks, P.A., Inhibition of [^{3}H]dopamine uptake into rat striatal slices by quaternary *N*-methylated nicotine metabolites, *Life Sci.*, 50, PL-223, 1992.

199. Yamashita, H., Kitayama, S., Zhang, Y.X., Takahashi, T., Dohi, T., and Nakamura, S., Effect of nicotine on dopamine uptake in COS cells possessing the rat dopamine transporter and in PC12 cells, *Biochem. Pharmacol.*, 49, 742, 1995.

200. Lerner-Marmarosh, N., Carroll, F.I., and Abood, L.G., Antagonism of nicotine's action by cocaine analogs, *Life Sci.*, 56, PL67, 1995.

201. Drew, A.E., Derbez, A.E., and Werling, L.L., Nicotinic receptor-mediated regulation of transporter activity in the rat prefrontal cortex, *Synapse,* 38, 10, 2000.

202. Drew, A.E. and Werling, L.L., Nicotinic receptor-mediated regulation of the dopamine transporter in rat prefrontocortical slices following chronic *in vivo* administration of nicotine, *Schizophr. Res.,* 65, 47, 2003.

203. Middleton, L.S., Cass, W.A., and Dwoskin, L.P., Nicotinic receptor modulation of dopamine transporter function in rat striatum and medial prefrontal cortex, *J. Pharmacol. Exp. Ther.*, 308, 367, 2003.

204. Li, S., Kim, K.Y., Kim, J.H., Kim, J.H., Park, M.S., Bahk, J.Y., and Kim, M.O., Chronic nicotine and smoking treatment increases dopamine transporter mRNA expression in the rat midbrain, *Neurosci. Lett.,* 363, 29, 2004.

205. Harrod, S.B., Mactutus, C.F., Bennett, K., Hasselrot, U., Wu, G., Welch, M., and Booze, R.M., Sex differences and repeated intravenous nicotine: behavioral sensitization and dopamine receptors, *Pharmacol. Biochem. Behav.*, 78, 581, 2004.

206. Batra, V., Patkar, A.A., Berrettini, W.H., Weinstein, S.P., and Leone, F.T., The genetic determinants of smoking, *Chest*, 123, 1730, 2003.

207. Arinami, T., Ishiguro, H., and Onaivi, E.S., Polymorphism in genes involved in neurotransmission in relation to smoking, *Eur. J. Pharmacol.*, 410, 221, 2000.

208. Vandenbergh, D.J., Persico, A.M., Hawkins, A.L., Griffin, C.A., Li, X., Jabs, E.W., et al., Human dopamine transporter gene (DAT1) maps to chromosome 5p15.3 and displays VNTR, *Genomics,* 14, 1104, 1992.

209. Fuke, S., Suo, S., Takahashi, N., Koike, H., Sasagawa, N., and Ishuiri, S., The VNTR polymorphism of the human dopamine transporter (DAT1) gene affects gene expression, *Pharmacogenom. J.*, 1, 152, 2001.

210. Comings, D.E. and Blum, K., Reward deficiency syndrome: genetic aspects of behavioral disorders, *Prog. Brain Res.,* 126, 325, 2000.

211. Gelernter, J., Kranzler, H.R., Satel, S.L., and Rao, P.A., Genetic association between dopamine transporter protein alleles and cocaine induced paranoia, *Neuropsychopharmacology,* 11, 195, 1994.

212. Lott, D.C., Kim, S., Cook, E.H., and de Wit, H., Dopamine transporter gene associated with diminished subjective response to amphetamine, *Neuropharmacology,* 1, 2004.

213. Schuckit, M.A., Low level of response to alcohol as predictor of alcoholism, *Am. J. Psychol.*, 151, 184, 1994.

214. Liu, H., Lin, S., Liu, S., Chen, S., Hu, C., Chang, J., and Leu, S., DAT polymorphism and diverse clinical manifestations in methamphetamine abusers, *Psychiatr. Gen.*, 14, 33, 2004.

215. Kreek, M.J., Bart, G., Lilly, C., LaForge, K.S., and Nielsen, D.A., Pharmacogenetic and human molecular genetics of opiate and cocaine addictions and their treatments, *Pharmacol. Rev.*, 57, 1, 2005.

216. Galeeva, A.R., Greeva, A.E., Yur'ev, E.B., and Khusnutdinova, E.K., VNTR polymorphism of the serotonin transporter and dopamine transporter genes in male opiate addicts, *Mol. Biol.*, 36, 462, 2002.

217. Foley, P.F., Loh, E.W., Innes, D.J., Williams, S.M., Tannenberg, A.E., Harper, C.G., and Dodd, P.R., Association studies of neurotransmitter gene polymorphisms in alcoholic Caucasians, *Ann. N.Y. Acad. Sci.*, 1025, 39, 2004.

218. Gorwood, P., Limosin, F., Batel, P., Hamon, M., Ades, J., and Boni, C., The A9 allele of the dopamine transporter gene is associated with delirium tremens and alcohol-withdrawal seizure, *Biol. Psychiatry,* 53, 85, 2003.

219. Kohnke, M.D., Batra, A., Kolb, W., Kohnke, A.M., Lutz, U., Schick, S., and Gaertner, I., Association of the dopamine transporter gene with alcoholism, *Alcohol*, 40(5), 339, 2005.

220. Jorm, A.F., Henderson, A.S., Jacob, P.A., Christensen, H., Korten, A. E., Rodgers, B., Tan, X., and Easteal, S., Association of smoking and personality with polymorphism of the dopamine transporter gene: results from a community survey, *Am. J. Med. Gen.*, 96, 331, 2000.

221. Lerman, C., Caporaso, N.E., Audrain, J., Main, D., Bowman, E.D., Lockshin, B., Boyd, N.R., and Shields, P.G., Evidence suggesting the role of specific genetic factors in cigarette smoking, *Health Psychol.*, 18, 14, 1999.

222. Sabol, S.Z., Nelson, M.L., Fisher, C., Gunzerath, L., Brody, C.L., Hu, S., Sirota, L.A., Marcus, S.E., Greenberg, B.D., Lucas, F.R., IV, Benjamin, J., Murphy, D.L., and Hamer, D.H., A genetic association for cigarette smoking behavior, *Health Psychol.*, 18, 7, 1999.

223. Ling, D., Niu, T., Feng, Y., Xing, H., and Xu, X., Association between polymorphism of the dopamine transporter gene and early smoking onset: an interaction risk on nicotine dependence, *J. Hum. Genet.*, 49, 35, 2004.

224. Deneau, G., Yanagita, T., and Seevers, M.H., Self-administration of psychoactive substances by the monkey, *Psychopharmacologia*, 16, 30, 1969.
225. Beardsley, P.M., Balster, R.L., and Harris, L.S., Self-administration of methylendioxymethamphetamine (MDMA) by rhesus monkeys, *Drug Alcohol Depend.*, 18, 149, 1986.
226. Spampinato, U., Espisito, E., and Samainin, R., Serotonin agonists reduce dopamine synthesis in the striatum only when the impulse flow of nigro-striatal neurons is intact, *J. Neurochem.*, 45, 980, 1985.
227. Hetey, L., Schwitzlowsky, R., and Oelssner, W., Influence of psychotomimetics and lisuride on synaptosomal dopamine release in the nucleus accumbens of rats, *Eur. J. Pharmacol.*, 93, 213, 1983.
228. Hetey, L. and Quirling, K., Synaptosomal uptake and release of dopamine and 5-hydroxytryptamine in the nucleus accumbens *in vitro* following *in vivo* administration of lysergic acid diethlamide in rats, *Acta Biol. Med. Ger.*, 39, 889, 1980.
229. Ator, N.A. and Ator, R.R., Self-administration of barbiturates and benzodiazepines: a review, *Pharmacol. Biochem. Behav.*, 27, 391, 1987.
230. Murai, T., Koshikawa, N., Kanayama, T., Takada, K., Tomiyama, K., and Kobayashi, M., Opposite effects of midazolam and beta-carboline-3-carboxylate ethyl ester on the release of dopamine from rat nucleus accumbens measured by *in vivo* microdialysis, *Eur. J. Pharmacol.*, 261, 65, 1994.
231. Finlay, J.M., Damsma, G., and Fibiger, H.C., Benzodiazepine-induced decreases in extracellular concentration of dopamine in the nucleus accumbens after acute and repeated administration, *Psychopharmacology*, 106, 202, 1992.
232. Louilot, A., Le Moal, M., and Simon, H., Presynaptic control of dopamine metabolism in the nucleus accumbens. Lack of effect of buspirone as demonstrated using *in vivo* voltammetry, *Life Sci.*, 40, 2017, 1987.
233. Reith, M.E.A., Sershen, H., and Lajtha, A., effects of caffeine on monoaminergic systems in mouse brain, *Acta Biochem. Biophys. Hung.*, 22, 149, 1987.
234. Corrigall, W.A. and Coen, K.M., Nicotine maintains robust self-administration in rats on a limited access schedule, *Psychopharmacology*, 99, 473, 1989.
235. Balster, R.L., Johanson, C.E., Harris, R.T., and Schuster, C.R., Phencyclidine self-administration in the rhesus monkey, *Pharmacol. Biochem. Behav.*, 1, 167, 1973.
236. Robinson, T.E. and Berridge, K.C., The neural basis of drug craving: An incentive-sensitization theory of addiction, *Brain Res. Rev.*, 18, 247, 1993.
237. Koob, G.F. and Le Moal, M., Drug abuse: Hedonic homeostatic dysregulation, *Science*, 278, 52, 1997.
238. Volkow, N.D., Wang, G.-J., Fowler, J.S., Gatley, S.J., Logan, J., Ding, Y.-S., Hitzeman, R., and Pappas, N., Dopamine transporter occupancies in the human brain induced by therapeutic doses of oral methylphenidate, *Am. J. Psychiatry*, 155, 1325, 1998.
239. Volkow, N.D., Wang, G.-J., Fischman, M.W., Foltin, R., Fowler, J.S., Franceschi, D., Fraceschi, M., Logan, J., Gatley, S.J., Wong, C., Ding, Y.-S., Hitzeman, R., and Pappas, N., Effects of route of administration on cocaine induced dopamine transporter blockade in the human brain, *Life Sci.*, 67, 1507, 2000.
240. Volkow, N.D., Wang, G.-J., Fowler, J.S., Logan, J., Gatley, S.J., Wong, C., Hitzeman, R., and Pappas, N., Reinforcing effects of psychostimulants in humans are associated with increases in brain dopamine and occupancy of D^2 receptors, *J. Pharmacol. Exp. Ther.*, 291, 409, 1999.
241. Volkow, N.D., Wang, G.-J., Fowler, J.S., Gatley, S.J., Logan, J., Ding, Y.-S., Dewey, S.L., Hitzeman, R., Gifford, A.N., and Pappas, N., Blockade of striatal dopamine transporters by intravenous methylphenidate is not sufficient to induce self-reports of "high," *J. Pharmacol. Exp. Ther.*, 288, 14, 1999.
242. Shimada, S., Kitayama, S., Lin, C.-L., Patel, A., Nathankumar, E., Gregor, P., Kuhar, M.J., and Uhl, G., Cloning and expression of a cocaine-sensitive dopamine transporter complementary DNA, *Science*, 254, 576, 1991.
243. Amara, S. and Kuhar, M.J., Neurotransmitter transporters: recent progress, *Annu. Rev. Neurosci.*, 16, 73, 1993.
244. Giros, B. and Caron, M.G., Molecular characterization of the dopamine transporter, *TIPS*, 14, 43, 1993.

CHAPTER 2

Neurochemistry of Nicotine Dependence

Darlene H. Brunzell, Ph.D.
Department of Psychiatry, Yale University School of Medicine, New Haven, Connecticut

CONTENTS

Tobacco use is the leading preventable cause of death in North America and a growing medical problem in developing countries throughout the world. In the Western world, the rising cost of cigarettes, social mores, and public policy against smoking have led to appreciable decreases in cigarette use over the last 25 years.[1,2] In recent years, however, smoking prevalence has appeared to reach asymptote at approximately 25%.[3,4] Those with schizophrenia, a history of depression, alcoholism or polydrug use, and those who have difficulty quitting with the help of currently available cessation methods continue to smoke.[3,5] Until recently, there were only two FDA-approved treatments for tobacco cessation: nicotine replacement therapy and bupropion. In May 2006, the FDA approved the use of a nicotinic receptor partial agonist, varenicline, for treatment of tobacco dependence. Whereas these therapies have realized some success, there remains an apparent need for novel treatments for nicotine and tobacco dependence. Nicotine is believed to be a major psychoactive component in cigarettes and smokeless tobacco. Advancing our under-standing of the neurochemical mechanisms of nicotine use and how nicotine-associated changes in neurochemistry relate to behaviors that support addiction will not only lead to novel treatments for tobacco cessation, but might also lead to advanced therapies for diseases that have high comorbidity with tobacco use. This chapter reviews nicotinic receptor composition, followed by a systems overview of how various nicotinic receptor subtypes are thought to contribute to nicotine reinforcement and incentive motivational processes. Because nicotine dependence is thought to reflect changes in communication between areas of the brain that control motivation, cognition, and reward, candidate intracellular signaling proteins thought to promote nicotine-dependent neuroplasticity are discussed, and finally the promise of novel compounds for tobacco cessation and their potential clinical applications are discussed.

2.1 NICOTINIC RECEPTOR COMPOSITION

Nicotine action is mediated through the nicotinic acetylcholine receptors (nAChRs). Although slightly different in subunit composition, most of our notions about neuronal nAChR structure and function are derived from exquisite work on nAChRs in the torpedo electric organ and at the neuromuscular junction (for detailed review, see References 6 through 9). Members of the ligand-gated superfamily of receptors, nAChRs respond endogenously to acetylcholine (ACh) in the periphery and central nervous system (CNS).[6] There are two general classes of nAChRs in the brain, both pentameric in structure. Neuronal nAChRs are either heteropentameres, made up of a combination of five α_2–α_6 and β_2–β_4 receptor subunits, or are homomeric in structure, made up of five α_7 subunits (Figure 2.1). Each subunit contains an N-terminal agonist binding domain, four transmembrane domains (M1 to M4), a large cytoplasmic loop between M3 and M4, and an extracellular C terminus.[10,11] The nAChRs exist in a variety of functional states including a closed, resting state, an open, activated state, a desensitized, unresponsive state, and an irreversible, inactive state.[12] When activated, the M2 domain of the nAChR undergoes a conformational change making the ion pore of the receptor permeable to cations (e.g., Na$^+$ and Ca^{2+};[10,13,14]) that lead to cellular activation, modification of second messenger signaling, and enhancement of neurotransmitter release.

The nAChR subtypes vary in response to pharmacological manipulation. The α_7 receptors have a low affinity for nicotine and are sensitive to α-bungarotoxin (α-BTX) antagonism, whereas the heteromeric nAChRs are not.[14] The β_2 containing (β_2*: asterisk denotes the presence of additional subunits) nAChRs have the highest affinity for nicotine binding and some selectivity for antagonism

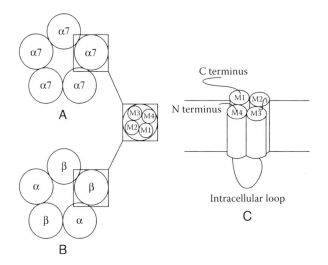

Figure 2.1   Diagram of nicotinic acetylcholine receptor (nAChR) structure. A top view of (A) an α_7 nAChR and (B) a β_2*nAChR shows that homomeric and heteromeric classes of nAChRs are both pentameric in structure. Each subunit is made up of four transmembrane domains with the M2 domain making up the ion pore. (C) A side view of the four transmembrane regions shows the N terminus, C terminus, and large M3–M4 intracellular loop that make up each nAChR subunit. The extracellular loops are available for binding to ligands and the intracellular loop is available for regulation of the nAChR by intracellular signaling proteins.

with dihydro-beta-erythroidine (DHβE),[15] and the α_3^* and α_6^*nAChRs are the only subtypes known to respond to α-conotoxin MII.[16–21] After some period of nAChR stimulation, there are conformational changes in the receptors[22,23] that cause them to become transiently unavailable for activation by nicotinic agonists,[24] sometimes irreversibly.[25] This desensitization of the receptors is thought to be regulated by calcium-mediated protein kinases at the intracellular loop between M3 and M4,[22,26] providing negative feedback to the nAChRs. The variability in sequence homology between nAChR subtypes at the intracellular loop may be responsible for the different rates of desensitization identified for the α_7 and β_2^*nAChRs.[27–29] Once bound by acetylcholine or nicotinic agonists, nAChR effects on neurochemistry depend on the conformation of the receptor, neuroanatomical localization of the receptor subtype, and the intracellular consequences of nAChR activation.

2.2 NEUROCHEMICAL SYSTEMS THAT SUPPORT NICOTINE USE

The prevailing belief in the drug addiction field is that with repeated drug use, neuroplasticity occurs within areas of the brain that modulate motivation, impulsivity, and reward.[30–32] These neurochemical changes are thought to support addictive behaviors and to transform the non-addicted brain into an addicted one. Much of the animal work to date has focused on the neurochemical mechanisms of nicotine reinforcement. Drug reinforcement is not included in the DSM-IV addiction criteria for good reason. A person can enjoy the pleasurable properties of a glass of wine without having any particular risk for alcoholism. If a drug such as nicotine is not positively or negatively reinforcing, however, it will not be sufficiently administered in order for nicotine dependence to develop. In this context, understanding the mechanisms of nicotine reinforcement might help identify genetic vulnerabilities for or protection from developing an addictive phenotype.[33] *Nicotine dependence* is a much more complex behavioral phenomenon. Following repeated use, incentive motivational processes (e.g., craving) come to regulate drug intake even in the absence of drug reinforcement or relief of symptoms of withdrawal.[34,35] Repeated association of cues with a primary reinforcer, such as nicotine, results in the ability of those cues to reinforce behaviors like drug seeking.[16]

2.2.1 Nicotine Reinforcement

2.2.1.1 *The Mesocorticolimbic Dopamine System*

Like other drugs of abuse, the reinforcing effects of nicotine are modulated, in large part, via the mesocorticolimbic dopamine (DA) system. Animal studies have shown that systemic and ventral tegmental area (VTA) administration of nicotine results in DA release to the nucleus accumbens (NAc).[36–38] Accumbens DA release increases with repeated nicotine exposure.[36] This neuroplasticity, termed sensitization, coincides with nicotine reinforcement[39–41] and locomotor activating effects of nicotine.[36,37] Both blockade of VTA nicotinic receptors[42,43] and destruction of DA inputs to the NAc[44] greatly reduce nicotine self-administration and conditioned place preference (CPP)† in rats. Unlike other psychostimulants, which enhance dopamine release via binding to dopamine transporters, nicotine regulation of dopamine is less direct. Although much evidence suggests that nAChRs act postsynaptically to enhance DA neuron activity,[45,46] emerging evidence indicates that VTA and NAc nAChRs act presynaptically to modulate neurotransmitter release[19,28,47] and regulate transporter function.[48]

---

† Conditioned place preference refers to a Pavlovian learning paradigm in which animals are repeatedly exposed to two novel adjacent chambers, one paired with nicotine administration and the other paired with saline injection. During the test the animal is allowed to cross between compartments. An increased amount of time spent in the drug-paired chamber is thought to reflect drug reinforcement and is defined as conditioned place preference.

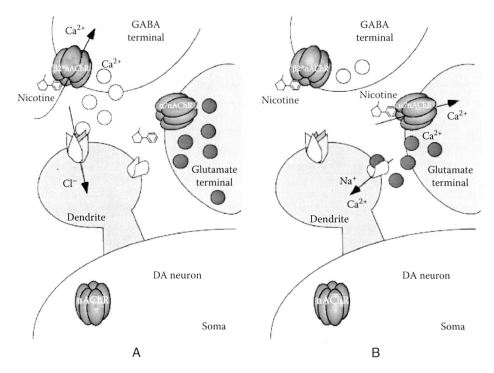

Figure 2.2   A presynaptic model of nicotine stimulation of ventral tegmental area DA neurons. (A) Nicotine first binds to the high-affinity β_2 containing nicotinic acetylcholine receptors (β_2*nAChRs), which reside on neuron terminals that release the inhibitory neurotransmitter GABA. Entry of calcium (Ca^{2+}) through the nAChR ion pore facilitates vesicle docking and neurotransmitter release. (B) The inhibitory GABA input to the DA neurons is short-lived, however, due to a fast desensitization of the β_2*nAChRs. As nicotine accumulates, it binds to the lower-affinity α_7 nAChRs that reside on the terminals of neurons that release the excitatory neurotransmitter, glutamate. Together nAChR-regulated disinhibition of GABA input and stimulation of glutamate input result in a net elevation of DA neuron activity and DA release in VTA projection areas.

An accumulation of data suggests that both the β_2* and α_7 receptor subtypes contribute to nicotine-induced increases in DA release and associated nicotine-dependent behaviors.[28,39,40,42,43,49,50] In the VTA, α_7 and β_2*nAChRs, respectively, reside on glutamatergic and GABAergic terminals. Electrophysiological data indicate that the higher affinity β_2*nAChRs are the first to be activated by nicotine (Figure 2.2A). In the VTA slice preparation, the β_2*nAChRs desensitize very quickly, becoming inactivated.[28,47] Because β_2*nAChRs stimulate γ-aminobutyric acid (GABA) release, desensitization of these receptors results in disinhibition of VTA DA neurons. Removal of GABA release on DA neurons is coincident with activation of the lower-affinity α_7 nAChRs, which facilitate excitatory glutamatergic input to the DA neurons (Figure 2.2B), resulting in a net increase in DA neuron firing.[28] At the DA terminals, however, β_2*nAChRs ($\alpha_4\beta_2$, $\alpha_6\beta_3\beta_2$, $\alpha_4\alpha_6\beta_3\beta_2$, $\alpha_4\alpha_5\beta_2$) and not α_7 nAChRs support nicotine-stimulated DA release.[19]

Studies in knockout mice indicate that the β_2*nAChRs are necessary for nicotine self-administration, DA-dependent locomotor activation, and nicotine-associated enhancement of NAc DA release.[40,51–53] Combined with studies showing that antagonism of the high-affinity nAChRs block self-administration,[44,54] it would appear that β_2*nAChRs are particularly critical for nicotine reinforcement. Unlike wild-type mice that self-administer both cocaine and nicotine, β_2*nAChR-null mutant mice learn to self-administer cocaine normally, but stop bar pressing as though receiving saline when cocaine is switched to nicotine.[40] Self-administration of VTA nicotine and associated DA release is rescued, however, in β_2*nAChR knockout mice with lentiviral-mediated expression of β_2 subunit DNA in the VTA.[55] Whereas several configurations of the β_2*nAChRs exist at the level of the VTA, much data point to the $\alpha_4\beta_2$ nicotinic receptors as playing a primary role in nicotine reinforcement. Mice lacking the α_4*nAChRs fail to show nicotine-dependent enhancements of DA release,[53] and a single nucleotide leucine-to-alanine α_4 mutation in the pore-forming M2 domain renders the α_4*nAChRs hypersensitive to nicotine stimulation and promotes conditioned place preference at otherwise sub-optimal doses of nicotine.[56] Together, these data suggest that the β_2*nAChRs are necessary and the α_4*nAChRs are sufficient for nicotine reinforcement. Interestingly, the α_4*nAChR knockout animals but not the β_2-null mutant mice show an increase in basal DA release to the NAc,[40,53] indicating that receptor conformations in addition to $\alpha_4\beta_2$ mediate DA input to the NAc.

Another candidate receptor subunit for nicotine reinforcement that has been less studied is α_6. The α_6 subunit associates with the β_2, β_3, and α_4 nAChR subunits in the CNS.[19,20,57,58] Unlike $\alpha_4\beta_2$ nAChRs, which are ubiquitously expressed throughout the brain, α_6 mRNA is chiefly expressed in catecholaminergic nuclei,[58] with receptor expression on DA terminals in the striatum.[59] Although no direct link has been made regarding the role of this receptor subunit in nicotine reinforcement, α_6 is well suited to contribute to neuroplasticity associated with nicotine exposure. α_6*nAChRs are capable of modulating nicotine-associated DA release at striatal DA terminals[19,57] and are upregulated following chronic nicotine exposure,[60] suggesting that the α_6 subunit might contribute to nicotine-associated changes in DA release that correlate with locomotor activation and nicotine reinforcement.

As α_7 nAChRs are known to reside on glutamate terminals in the VTA,[61] the role of α_7 nAChRs in nicotine-elicited dopamine release is supported by studies that manipulate glutamate receptor function. Glutamate receptor antagonism in the VTA greatly reduces nicotine-associated increases in NAc DA release without affecting baseline levels of accumbens DA.[62] Behaviorally, NMDA glutamate receptor antagonism blocks nicotine locomotor sensitization in rats.[63] As the reports of α_7 antagonism on nicotine reinforcement are equivocal,[42,54,64] it is unclear what role the α_7 nAChRs play in nicotine reward. Local administration of 4 nM methyllycaconitine (MLA) into the VTA reverses nicotine-conditioned place preference,[42] and high doses of this putatively selective α_7 antagonist (3.9 and 7.8 mg/kg i.p.) attenuate nicotine self-administration in rats, suggesting that α_7 nAChRs contribute to nicotine reinforcement.[64] Similar doses of MLA achieved in brain,[65] however, block nicotine-stimulated DA release in striatal synaptosome preparations that do not contain α_7 nicotinic receptors,[19,66] bringing the selectivity of MLA for α_7 nAChRs into question at higher doses.[66] The fact that MLA blocks α conotoxin MII binding at behaviorally efficacious doses[20,67] raises the possibility that antagonism of α_3* or α_6*nAChRs in addition to α_7 nAChRs might be responsible for MLA-dependent attenuation of nicotine reinforcement.

2.2.1.2 Hindbrain Inputs to the VTA

Hindbrain regions including the pedunculopontine tegmental nucleus (PPT) and lateral dorsal tegmental nucleus (LDT) give rise to acetylcholinergic, GABAergic, and glutamatergic projections to the VTA that are thought to regulate drug reward.[68–70] Local infusion of GABA receptor agonists and lesions to the PPT result in a marked attenuation of nicotine-associated locomotor activation, nicotine CPP, and nicotine self-administration in rodents.[71–73] PPT administration of DHβE also greatly attenuates nicotine self-administration in rats,[72] suggesting that PPT-regulated nicotine reinforcement is mediated in part by high-affinity β_2*nAChRs. Nicotinic receptor antagonism also inhibits ACh release in PPT synaptosome preparation.[67]

Various studies suggest that basal forebrain cholinergic projections and accumbens ACh interneurons may also regulate behavior associated with the reinforcing properties of cocaine, morphine, and ethanol.[74–78] Whereas muscarinic ACh receptors might also meter behaviors associated with drug reinforcement, studies show that nAChR stimulation enhances and antagonism attenuates cocaine CPP. β_2-null mutant mice are also slightly impaired at cocaine CPP.[79] Given that ACh appears to modulate both drug aversion and reward,[42,76] it is possible that nAChRs in mesolimbic DA areas regulate motivational valence or learning and memory processes that underlie drug use and not drug reinforcement per se. There is very high comorbidity for tobacco use with substance use disorders.[3] The specific contributions of nAChRs to drug reinforcement, more broadly defined, remain to be determined.

2.2.1.3   Beyond the Role of DA in Nicotine Reinforcement

Although the research described thus far supports the tenet that nicotine reinforcement is regulated by the ability of nAChRs to enhance mesolimbic DA release, an accumulation of evidence questions the simplicity of this dogma. Despite treatment with neuroleptics that block DA receptor stimulation, the percentage of people with schizophrenia who smoke is several times greater than the population as a whole.[3,5] In rats, the effects of intra-VTA infusion of nicotine on behavior are dose dependent; animals display conditioned place aversion at low doses and CPP at steadily increasing doses of nicotine.[80] The experimenters found that intra-accumbens and systemic administration of the neuroleptic, α-flupenthixol, reversed the conditioned aversive but not rewarding effects of nicotine, concluding that NAc dopamine regulates nicotine aversion and not reward.[80] α-Flupenthixol, however, blocks both Gs-coupled, D_1- and Gi-coupled, D_2-type DA receptors, which are known to have opposite effects on the cAMP signaling pathway (Figure 2.3).[31] Recent evidence suggests that cAMP-responsive element-binding protein regulates both rewarding and aversive effects of morphine.[81] Together these data suggest that NAc DA and the cAMP pathway might serve to regulate motivational valence rather than drug reinforcement per se.

Electrophysiological data show that while pulses of ACh enhance DA neuron activity as one might expect with acute nicotine exposure, simulation of steady states of human nicotine concentrations[82] quickly results in desensitization of the midbrain nAChRs.[47] Indeed, striatal synaptosome preparation used to measure DA release shows that much lower doses of nicotine are required for desensitization than for activation of nAChRs.[24,83] This acute tolerance might account for smoker reports that the first cigarette of the day is most pleasurable.[84] In human brain, β_2*nAChR binding is prolonged for as long as 5 h after a smoking episode,[85] begging the question as to why people continue to smoke throughout the day. Research using electrochemical cyclic voltammetry shows that nAChR regulation of DA release depends upon the state of the DA neuron during nicotine application.[86,87] When DA neurons are held in a tonic or "resting" state, nicotine decreases DA release, but when DA neurons are in a phasic state, as one would expect during the presentation of a reward,[88] nicotine enhances DA release.[86] Interestingly, DA neurons respond similarly to nicotine and nAChR antagonists, suggesting that nicotine's action on DA release is mediated by desensitization of the receptor.[86,87] Over time, cues come to elicit phasic activity of DA neurons where primary reinforcers once did.[88] These data may explain at an electrophysiological level how cigarette-associated cues maintain smoking behavior.

2.2.2   Neurochemistry of Cue-Driven Behaviors

Although the NAc has received the most attention for its role in nicotine reinforcement, other VTA projection areas including the hippocampus, prefrontal cortex, and amygdala contribute to the control that cues have over behavior, or conditioned reinforcement.[30,32] Such behaviors may represent changes in incentive motivation that perpetuate drug use even in the absence of drug reinforcement.[34] Sensory cues associated with the act of inhaling regulate the degree to which smokers find pleasure in smoking denicotinized cigarettes.[89,90] The VTA, NAc, amygdala, and prefrontal cortex are activated in humans during craving and the presentation of cigarette-associated cues,[91,92] indicating that these areas of the brain contribute to conditioned reinforcement associated with cigarette smoking.

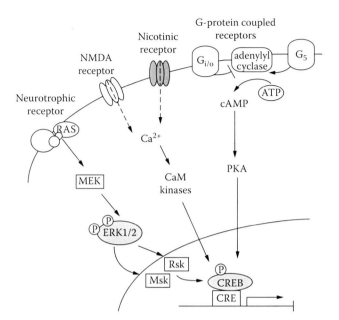

Figure 2.3 Mechanisms by which nicotine might affect ERK and CREB signaling. Nicotine stimulation of glutamate release or direct activation of nicotinic acetylcholine receptors (nAChRs) results in the influx of calcium (Ca^{2+}), among other cations, through NMDA glutamate and nAChRs. Intracellular Ca^{2+} can result in activation of Ca^{2+}/calmodulin-dependent protein kinases that lead to phosphorylation and activation of the transcription factor, cAMP-responsive element binding protein (CREB). Nicotine-associated changes in levels of growth factors result in changes in activation of neurotrophic receptors that stimulate extracellular regulated protein kinase (ERK) and downstream activation of CREB via protein kinases, ribosomal S6 kinase (Rsk), and mitogen- and stress-activated protein kinase (Msk). *In vitro* studies show that fast activation of ERK by nicotine is Ca^{2+}-dependent and mediated via voltage-gated Ca^{2+} channels;[119,120] however, the intracellular mechanism of Ca^{2+}-mediated ERK activation remains to be determined. Nicotine-stimulated elevations of DA release can lead to activation of G protein-coupled receptors, which in turn modify cAMP signaling and downstream activation of protein kinase A (PKA), a kinase known to directly phosphorylate CREB and promote CRE-mediated transcription.

Animal studies have shown that cues greatly enhance the degree to which animals will self-administer nicotine[34,93–95] and can support self-administration behavior for weeks after the removal of nicotine.[93,96] In rats, a nicotine-associated cue is a more efficient primer than nicotine itself at reinstating self-administration,[97] and a nicotine-paired context can elicit changes in immediate early gene activity in the prefrontal cortex,[98] suggesting that conditioned reinforcement for nicotine-associated cues occurs at a molecular level. Like other drugs of abuse, the control of nicotine-associated cues over behavior is likely mediated within areas of the brain that receive DA and glutamate stimulation.[32] One theory suggests that coincident activation of NAc neurons by DA and glutamate supports drug reinforcement and natural reward.[99] Blockade of metabotropic glutamate receptor 5 ($mGluR_5$) with the antagonist MPEP not only decreases nicotine self-administration and break points for nicotine,[100,101] but also significantly attenuates cue-induced reinstatement of nicotine self-administration.[102] Disruption of D_3 DA receptors, which are upregulated with repeated nicotine exposure,[103] significantly attenuates behavioral locomotor sensitization in response to a nicotine-paired context.[104] D_3 antagonists and partial agonists also block nicotine-conditioned place preference[105] suggesting that manipulation of D_3 receptors might be efficacious in reducing nicotine seeking or nicotine reinforcement. The efficacy of D_3 partial agonists and antagonists in blocking nicotine self-administration remains to be tested, however.

Not only do cues control nicotine use, but nicotine exposure also enhances conditioned reinforcement in rats and mice for weeks following exposure to nicotine[106–109] (Figure 2.4), and can

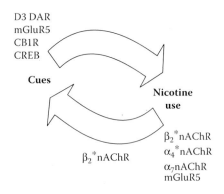

Figure 2.4   A perpetual learning model for nicotine dependence. Evidence shows that cues greatly enhance nicotine self-administration and that nicotine exposure augments conditioned reinforcement for natural and drug reinforcers. Although drug reinforcement does not necessarily lead to addiction, nicotine reinforcement most likely facilitates the development of nicotine dependence. Evidence suggests that the β_2^*, α_4^*, and α_7 nicotinic acetylcholine receptors (nAChRs) and metabotropic glutamate receptor 5 (mGluR5) glutamate receptors contribute to nicotine self-administration. The D_3 dopamine receptors (D3 DAR), CB_1 cannabinoid receptors (CB1R), mGluR5 glutamate receptors, and the transcription factor CREB appear to be involved in cue-associated changes in neuroplasticity and the control of nicotine-dependent behaviors. Nicotine-associated enhancement of conditioned reinforcement for cues paired with a natural reinforcer requires β_2^*nAChRs. β_2^*nAChRs might also serve to amplify the conditioned reinforcement properties of nicotine-associated cues.

act as an occasion setter to facilitate the association of cues with reward.[110] Studies in β_2-null mutant mice show that nicotine enhancement of conditioned reinforcement is dependent on the presence of the β_2^*nAChRs.[106] The cannabinoid receptor 1 (CB_1) antagonist, rimonabant, appears to curb both primary and incentive motivation processes affected by nicotine,[106] blocking control of conditioned reinforcers over nicotine intake and having potential to decrease weight gain associated with quit attempts.[111] Nicotine's ability to act as a primary reinforcer in addition to its ability to enhance learning and incentive motivational processes may explain why people and animals have difficulty abandoning behaviors associated with tobacco smoking and nicotine intake.

2.3 NICOTINE-ASSOCIATED CHANGES IN INTRACELLULAR SIGNALING

At the cellular level, nicotine-induced changes in second messenger signaling are thought to support nicotine-associated changes in neurochemistry and behavioral phenotypes. Due to their putative roles in cellular processes underlying learning and memory (for detailed review, see References 112 and113), the extracellular regulated protein kinase (ERK) and cyclic AMP responsive element binding (CREB) signaling pathways have received the most attention for their potential roles in neuroplasticity underlying nicotine dependence (Figure 2.3).[114–117] *In vitro* studies have shown that ERK is activated by nicotine treatment[118] and is critical for nicotine-dependent activation of CREB[119,120] and tyrosine hydroxylase, the rate-limiting enzyme in DA synthesis.[121,122] *In vivo* studies show that regulation of ERK by nicotine is region and treatment specific.[114,116] Although acute administration of nicotine elevates levels of phosphorylated ERK (pERK) in the amygdala and prefrontal cortex,[116] chronic exposure to doses of nicotine known to have relevance for neural plasticity and locomotor activation[52,123] results in elevation of pERK in the prefrontal cortex, but leads to significant decreases in levels of ERK and pERK in the amygdala.[114] Amygdala changes in ERK protein expression following repeated nicotine exposure may support conditioned reinforcement processes; however, the role of ERK signaling in incentive motivation remains to be explored.

An accumulation of evidence suggests that the transcription factor CREB regulates the rewarding properties of nicotine. Unlike their wild-type counterparts, mice with a targeted mutation of CREB (CREB$^{\alpha\delta}$) fail to show nicotine-conditioned place preference following four pairings of a novel chamber with nicotine.[117] In wild-type mice, acute and four repeated exposures to nicotine both resulted in elevated levels of VTA pCREB,[117] suggesting that activation of CREB in the VTA might regulate the primary reinforcing properties of nicotine. Interestingly, the nicotine-paired chamber was also capable of eliciting an increase in pCREB,[117] showing that the nicotine-paired environment became a conditioned reinforcer capable of controlling intracellular signaling associated with nicotine exposure. Chronic and acute nicotine exposure and nicotine withdrawal have been shown to affect phosphorylation of CREB in the NAc, PFC, VTA, and amygdala.[114,115,117] NAc levels of pCREB differ between acute paradigms, where little to no change is observed,[114,117,124] and chronic exposure where marked decreases in NAc pCREB are evident.[114] Similarly, increases of pCREB in the prefrontal cortex are specific to chronic nicotine exposure in mice[114] and are observed to decrease in rats following nicotine withdrawal,[115] suggesting that CREB in the NAc and prefrontal cortex might regulate some conditioned emotive properties of nicotine reward or withdrawal. Nicotine withdrawal can precipitate an episode of depression[125] and inhibition of NAc CREB has antidepressant-like effects in rats.[126] More studies need to be done to clarify the contributions of the prefrontal cortex and NAc CREB in complex behaviors that support nicotine dependence.

2.4 SUMMARY AND CLINICAL IMPLICATIONS

Nicotine dependence is a complex biobehavioral phenomenon that is likely regulated by cue-driven incentive motivational processes. As suggested by the work described here, antagonism at mGluR$_5$ glutamate, D$_3$ DA, CB$_1$ cannabinoid, and β_2*nAChRs might have particular promise for promoting nicotine cessation. Preliminary trials indicate that quit rates for β_2*nAChR partial agonist varenicline are twice that reported for more traditional therapies.[127] Preclinical evidence suggests that even greater nicotine cessation outcomes might be achieved if varenicline is used in combination with behavioral therapies. If administered using techniques that enable local control of expression, CREB and ERK might serve as effective molecular targets for gene therapy. Other novel nicotine-cessation treatments under consideration include those that reduce the function of mu opioid receptors in the brain. Evidence suggests that naltrexone, an opiate antagonist that has enjoyed some success as a treatment for alcohol cessation,[128] should be considered for "off-label" nicotine cessation use.[129–131] Mu opioid receptors in the VTA appear to promote nicotine reward[117] and may be one point of convergence for nicotine and alcohol abuse potential. Last, a nicotine vaccine that limits the bioavailability of nicotine in the brain has been shown to lead to significant reductions in nicotine intake in preclinical trials.[132]

Despite that a large number of smokers want to quit, few are able to do so with currently approved treatments for nicotine dependence. Among those who have particular difficulty quitting smoking are those who suffer from polydrug use, depression, and schizophrenia.[3,5] There is large individual variability in responsiveness to nicotine and reasons for smoking.[84] Parsing out the specific contributions of nAChRs and their downstream neurochemical targets to various behaviors that support nicotine dependence may lead to treatments for nicotine cessation that are effective in a broader spectrum of individuals.

REFERENCES

1. Mendez, D., Warner, K.E., Courant, P.N. Has smoking cessation ceased? Expected trends in the prevalence of smoking in the United States. *Am. J. Epidemiol.* 148:249, 1998.

2. Stegmayr, B., Eliasson, M., Rodu, B. The decline of smoking in northern Sweden. *Scand. J. Public Health.* 33:321, 2005.

3. Kalman, D., Morissette, S.B., George, T.P. Co-morbidity of smoking in patients with psychiatric and substance use disorders. *Am. J. Addict.* 14:106, 2005.

4. Weintraub, J.M., Hamilton, W.L. Trends in prevalence of current smoking, Massachusetts and states without tobacco control programmes, 1990 to 1999. *Tob. Control.* 11(Suppl. 2):ii8, 2002.

5. Leonard, S., Adler, L.E., Benhammou, K. et al. Smoking and mental illness. *Pharmacol. Biochem. Behav.* 70:561, 2001.

6. Le Novere, N., Changeux, J.P. Molecular evolution of the nicotinic acetylcholine receptor: an example of multigene family in excitable cells. *J. Mol. Evol.* 40:155, 1995.

7. Karlin, A. Emerging structure of the nicotinic acetylcholine receptors. *Nat. Rev. Neurosci.* 3:102, 2002.

8. Lindstrom, J.M. Nicotinic acetylcholine receptors of muscles and nerves: comparison of their structures, functional roles, and vulnerability to pathology. *Ann. N.Y. Acad. Sci.* 998:41, 2003.

9. Picciotto, M.R., Caldarone, B.J., Brunzell, D.H., Zachariou, V., Stevens, T.R., King, S.L. Neuronal nicotinic acetylcholine receptor subunit knockout mice: physiological and behavioral phenotypes and possible clinical implications. *Pharmacol. Ther.* 92:89, 2001.

10. Karlin, A., Akabas, M.H. Toward a structural basis for the function of nicotinic acetylcholine receptors and their cousins. *Neuron.* 15:1231, 1995.

11. Corringer, P.J., Le Novere, N., Changeux, J.P. Nicotinic receptors at the amino acid level. *Annu. Rev. Pharmacol. Toxicol.* 40:431, 2000.

12. Changeux, J.P., Devillers-Thiery, A., Chemouilli, P. Acetylcholine receptor: an allosteric protein. *Science.* 225:1335, 1984.

13. Leonard, R.J., Labarca, C.G., Charnet, P., Davidson, N., Lester, H.A. Evidence that the M2 membrane-spanning region lines the ion channel pore of the nicotinic receptor. *Science.* 242:1578, 1988.

14. Arias, H.R. Localization of agonist and competitive antagonist binding sites on nicotinic acetylcholine receptors. *Neurochem. Int.* 36:595, 2000.

15. Whiteaker, P., Marks, M.J., Grady, S.R. et al. Pharmacological and null mutation approaches reveal nicotinic receptor diversity. *Eur. J. Pharmacol.* 393:123, 2000.

16. Mackintosh, N. *The Psychology of Animal Learning.* Academic Press, New York, 1974.

17. Kulak, J.M., Nguyen, T.A., Olivera, B.M., McIntosh, J.M. alpha-Conotoxin MII blocks nicotine-stimulated dopamine release in rat striatal synaptosomes. *J. Neurosci.* 17:5263, 1997.

18. McIntosh, J.M., Azam, L., Staheli, S. et al. Analogs of alpha-conotoxin MII are selective for alpha6-containing nicotinic acetylcholine receptors. *Mol. Pharmacol.* 65:944, 2004.

19. Salminen, O., Murphy, K.L., McIntosh, J.M., et al. Subunit composition and pharmacology of two classes of striatal presynaptic nicotinic acetylcholine receptors mediating dopamine release in mice. *Mol. Pharmacol.* 65:1526, 2004.

20. Salminen, O., Whiteaker, P., Grady, S.R., Collins, A.C., McIntosh, J.M., Marks, M.J. The subunit composition and pharmacology of alpha-Conotoxin MII-binding nicotinic acetylcholine receptors studied by a novel membrane-binding assay. *Neuropharmacology.* 48:696, 2005.

21. Vailati, S., Moretti, M., Balestra, B., McIntosh, M., Clementi, F., Gotti, C. Beta3 subunit is present in different nicotinic receptor subtypes in chick retina. *Eur. J. Pharmacol.* 393:23, 2000.

22. Fenster, C.P., Beckman, M.L., Parker, J.C., et al. Regulation of alpha4beta2 nicotinic receptor desensitization by calcium and protein kinase C. *Mol. Pharmacol.* 55:432, 1999.

23. Fenster, C.P., Hicks, J.H., Beckman, M.L., Covernton, P.J., Quick, M.W., Lester, R.A. Desensitization of nicotinic receptors in the central nervous system. *Ann. N.Y. Acad. Sci.* 868:620, 1999.

24. Grady, S.R., Marks, M.J., Collins, A.C. Desensitization of nicotine-stimulated [3H]dopamine release from mouse striatal synaptosomes. *J. Neurochem.* 62:1390, 1994.

25. Lukas, R.J. Effects of chronic nicotinic ligand exposure on functional activity of nicotinic acetylcholine receptors expressed by cells of the PC12 rat pheochromocytoma or the TE671/RD human clonal line. *J. Neurochem.* 56:1134, 1991.

26. Huganir, R.L., Delcour, A.H., Greengard, P., Hess, G.P. Phosphorylation of the nicotinic acetylcholine receptor regulates its rate of desensitization. *Nature.* 321:774, 1986.

27. Dani, J.A., Radcliffe, K.A., Pidoplichko, V.I. Variations in desensitization of nicotinic acetylcholine receptors from hippocampus and midbrain dopamine areas. *Eur. J. Pharmacol.* 393:31, 2000.

28. Mansvelder, H.D., Keath, J.R., McGehee, D.S. Synaptic mechanisms underlie nicotine-induced excitability of brain reward areas. *Neuron.* 33:905, 2002.
29. Wooltorton, J.R., Pidoplichko, V.I., Broide, R.S., Dani, J.A. Differential desensitization and distribution of nicotinic acetylcholine receptor subtypes in midbrain dopamine areas. *J. Neurosci.* 23:3176, 2003.
30. Jentsch, J.D., Taylor, J.R. Impulsivity resulting from frontostriatal dysfunction in drug abuse: implications for the control of behavior by reward-related stimuli. *Psychopharmacology* (Berlin). 146:373, 1999.
31. Nestler, E.J. Molecular basis of long-term plasticity underlying addiction. *Nat. Rev. Neurosci.* 2:119, 2001.
32. Robbins, T.W., Everitt, B.J. Limbic-striatal memory systems and drug addiction. *Neurobiol. Learn. Mem.* 78:625, 2002.
33. Lerman, C., Patterson, F., Berrettini, W. Treating tobacco dependence: state of the science and new directions. *J. Clin. Oncol.* 23:311, 2005.
34. Robinson, T.E., Berridge, K.C. Addiction. *Annu. Rev. Psychol.* 54:25, 2003.
35. Robinson, T.E., Berridge, K.C. The neural basis of drug craving: an incentive-sensitization theory of addiction. *Brain Res. Brain Res. Rev.* 18:247, 1993.
36. Benwell, M.E., Balfour, D.J. The effects of acute and repeated nicotine treatment on nucleus accumbens dopamine and locomotor activity. *Br. J. Pharmacol.* 105:849, 1992.
37. Di Chiara, G., Imperato, A. Drugs abused by humans preferentially increase synaptic dopamine concentrations in the mesolimbic system of freely moving rats. *Proc. Natl. Acad. Sci. U.S.A.* 85:5274, 1988.
38. Ferrari, R., Le Novere, N., Picciotto, M.R., Changeux, J.P., Zoli, M. Acute and long-term changes in the mesolimbic dopamine pathway after systemic or local single nicotine injections. *Eur. J. Neurosci.* 15:1810, 2002.
39. Epping-Jordan, M.P., Picciotto, M.R., Changeux, J.P., Pich, E.M. Assessment of nicotinic acetylcholine receptor subunit contributions to nicotine self-administration in mutant mice. *Psychopharmacology* (Berlin). 147:25, 1999.
40. Picciotto, M.R., Zoli, M., Rimondini, R. et al. Acetylcholine receptors containing the beta2 subunit are involved in the reinforcing properties of nicotine. *Nature.* 391:173, 1998.
41. Shoaib, M., Stolerman, I.P., Kumar, R.C. Nicotine-induced place preferences following prior nicotine exposure in rats. *Psychopharmacology* (Berlin). 113:445, 1994.
42. Laviolette, S.R., van der Kooy, D. The motivational valence of nicotine in the rat ventral tegmental area is switched from rewarding to aversive following blockade of the alpha7-subunit-containing nicotinic acetylcholine receptor. *Psychopharmacology* (Berlin). 166:306, 2003.
43. Corrigall, W.A., Coen, K.M., Adamson, K.L. Self-administered nicotine activates the mesolimbic dopamine system through the ventral tegmental area. *Brain Res.* 653:278, 1994.
44. Corrigall, W.A., Franklin, K.B., Coen, K.M., Clarke, P.B. The mesolimbic dopaminergic system is implicated in the reinforcing effects of nicotine. *Psychopharmacology* (Berlin). 107:285, 1992.
45. Klink, R., de Kerchove d'Exaerde, A., Zoli, M., Changeux, J.P. Molecular and physiological diversity of nicotinic acetylcholine receptors in the midbrain dopaminergic nuclei. *J. Neurosci.* 21:1452, 2001.
46. Wu, J., George, A.A., Schroeder, K.M. et al. Electrophysiological, pharmacological, and molecular evidence for alpha7-nicotinic acetylcholine receptors in rat midbrain dopamine neurons. *J. Pharmacol. Exp. Ther.* 311:80, 2004.
47. Pidoplichko, V.I., DeBiasi, M., Williams, J.T., Dani, J.A. Nicotine activates and desensitizes midbrain dopamine neurons. *Nature.* 390:401, 1997.
48. Middleton, L.S., Cass, W.A., Dwoskin, L.P. Nicotinic receptor modulation of dopamine transporter function in rat striatum and medial prefrontal cortex. *J. Pharmacol. Exp. Ther.* 308:367, 2004.
49. Pidoplichko, V.I., Noguchi, J., Areola, O.O. et al. Nicotinic cholinergic synaptic mechanisms in the ventral tegmental area contribute to nicotine addiction. *Learn. Mem.* 11:60, 2004.
50. Shoaib, M., Benwell, M.E., Akbar, M.T., Stolerman, I.P., Balfour, D.J. Behavioural and neurochemical adaptations to nicotine in rats: influence of NMDA antagonists. *Br. J. Pharmacol.* 111:1073, 1994.
51. Epping-Jordan, M.P., Watkins, S.S., Koob, G.F., Markou, A. Dramatic decreases in brain reward function during nicotine withdrawal. *Nature.* 393:76, 1998.
52. King, S.L., Caldarone, B.J., Picciotto, M.R. Beta2-subunit-containing nicotinic acetylcholine receptors are critical for dopamine-dependent locomotor activation following repeated nicotine administration. *Neuropharmacology.* 47(Suppl. 1):132, 2004.

53. Marubio, L.M., Gardier, A.M., Durier, S. et al. Effects of nicotine in the dopaminergic system of mice lacking the alpha4 subunit of neuronal nicotinic acetylcholine receptors. *Eur. J. Neurosci.* 17:1329, 2003.

54. Grottick, A.J., Trube, G., Corrigall, W.A. et al. Evidence that nicotinic alpha(7) receptors are not involved in the hyperlocomotor and rewarding effects of nicotine. *J. Pharmacol. Exp. Ther.* 294:1112, 2000.

55. Maskos, U., Molles, B.E., Pons, S. et al. Nicotine reinforcement and cognition restored by targeted expression of nicotinic receptors. *Nature.* 436:103, 2005.

56. Tapper, A.R., McKinney, S.L., Nashmi, R. et al. Nicotine activation of alpha4* receptors: sufficient for reward, tolerance, and sensitization. *Science.* 306:1029, 2004.

57. Champtiaux, N., Gotti, C., Cordero-Erausquin, M. et al. Subunit composition of functional nicotinic receptors in dopaminergic neurons investigated with knock-out mice. *J. Neurosci.* 23:7820, 2003.

58. Grinevich, V.P., Letchworth, S.R., Lindenberger, K.A. et al. Heterologous expression of human {alpha}6{beta}4{beta}3{alpha}5 nicotinic acetylcholine receptors: binding properties consistent with their natural expression require quaternary subunit assembly including the {alpha}5 subunit. *J. Pharmacol. Exp. Ther.* 312:619, 2005.

59. McCallum, S.E., Parameswaran, N., Bordia, T., McIntosh, J.M., Grady, S.R., Quik, M. Decrease in {alpha}3*/{alpha}6* nicotinic receptors but not nicotine-evoked dopamine release in monkey brain after nigrostriatal damage. *Mol. Pharmacol.* 68:737, 2005.

60. Parker, S.L., Fu, Y., McAllen, K. et al. Up-regulation of brain nicotinic acetylcholine receptors in the rat during long-term self-administration of nicotine: disproportionate increase of the alpha6 subunit. *Mol. Pharmacol.* 65:611, 2004.

61. Wonnacott, S., Kaiser, S., Mogg, A., Soliakov, L., Jones, I.W. Presynaptic nicotinic receptors modulating dopamine release in the rat striatum. *Eur. J. Pharmacol.* 393:51, 2000.

62. Schilstrom, B., Nomikos, G.G., Nisell, M., Hertel, P., Svensson, T.H. *N*-Methyl-D-aspartate receptor antagonism in the ventral tegmental area diminishes the systemic nicotine-induced dopamine release in the nucleus accumbens. *Neuroscience.* 82:781, 1998.

63. Shoaib, M., Schindler, C.W., Goldberg, S.R., Pauly, J.R. Behavioural and biochemical adaptations to nicotine in rats: influence of MK801, an NMDA receptor antagonist. *Psychopharmacology* (Berlin). 134:121, 1997.

64. Markou, A., Paterson, N.E. The nicotinic antagonist methyllycaconitine has differential effects on nicotine self-administration and nicotine withdrawal in the rat. *Nicotine Tob. Res.* 3:361, 2001.

65. Turek, J.W., Kang, C.H., Campbell, J.E., Arneric, S.P., Sullivan, J.P. A sensitive technique for the detection of the alpha 7 neuronal nicotinic acetylcholine receptor antagonist, methyllycaconitine, in rat plasma and brain. *J. Neurosci. Methods.* 61:113, 1995.

66. Mogg, A.J., Whiteaker, P., McIntosh, J.M., Marks, M., Collins, A.C., Wonnacott, S. Methyllycaconitine is a potent antagonist of alpha-conotoxin-MII-sensitive presynaptic nicotinic acetylcholine receptors in rat striatum. *J. Pharmacol. Exp. Ther.* 302:197, 2002.

67. Grady, S.R., Meinerz, N.M., Cao, J. et al. Nicotinic agonists stimulate acetylcholine release from mouse interpeduncular nucleus: a function mediated by a different nAChR than dopamine release from striatum. *J. Neurochem.* 76:258, 2001.

68. Garzon, M., Vaughan, R.A., Uhl, G.R., Kuhar, M.J., Pickel, V.M. Cholinergic axon terminals in the ventral tegmental area target a subpopulation of neurons expressing low levels of the dopamine transporter. *J. Comp. Neurol.* 410:197, 1999.

69. Kalivas, P.W. Neurotransmitter regulation of dopamine neurons in the ventral tegmental area. *Brain Res. Brain Res. Rev.* 18:75, 1993.

70. Omelchenko, N., Sesack, S.R. Laterodorsal tegmental projections to identified cell populations in the rat ventral tegmental area. *J. Comp. Neurol.* 483:217, 2005.

71. Corrigall, W.A., Coen, K.M., Zhang, J., Adamson, K.L. GABA mechanisms in the pedunculopontine tegmental nucleus influence particular aspects of nicotine self-administration selectively in the rat. *Psychopharmacology* (Berlin). 158:190, 2001.

72. Lanca, A.J., Adamson, K.L., Coen, K.M., Chow, B.L., Corrigall, W.A. The pedunculopontine tegmental nucleus and the role of cholinergic neurons in nicotine self-administration in the rat: a correlative neuroanatomical and behavioral study. *Neuroscience.* 96:735, 2000.

73. Laviolette, S.R., Alexson, T.O., van der Kooy, D. Lesions of the tegmental pedunculopontine nucleus block the rewarding effects and reveal the aversive effects of nicotine in the ventral tegmental area. *J. Neurosci.* 22:8653, 2002.

74. Alcantara, A.A., Chen, V., Herring, B.E., Mendenhall, J.M., Berlanga, M.L. Localization of dopamine D2 receptors on cholinergic interneurons of the dorsal striatum and nucleus accumbens of the rat. *Brain Res.* 986:22, 2003.

75. Berlanga, M.L., Olsen, C.M., Chen, V. et al. Cholinergic interneurons of the nucleus accumbens and dorsal striatum are activated by the self-administration of cocaine. *Neuroscience.* 120:1149, 2003.

76. Hikida, T., Kitabatake, Y., Pastan, I., Nakanishi, S. Acetylcholine enhancement in the nucleus accumbens prevents addictive behaviors of cocaine and morphine. *Proc. Natl. Acad. Sci. U.S.A.* 100:6169, 2003.

77. Nestby, P., Vanderschuren, L.J., De Vries, T.J. et al. Ethanol, like psychostimulants and morphine, causes long-lasting hyperreactivity of dopamine and acetylcholine neurons of rat nucleus accumbens: possible role in behavioural sensitization. *Psychopharmacology* (Berlin). 133:69, 1997.

78. Smith, J.E., Co, C., Yin, X. et al. Involvement of cholinergic neuronal systems in intravenous cocaine self-administration. *Neurosci. Biobehav. Rev.* 27:841, 2004.

79. Zachariou, V., Caldarone, B.J., Weathers-Lowin, A. et al. Nicotine receptor inactivation decreases sensitivity to cocaine. *Neuropsychopharmacology.* 24:576, 2001.

80. Laviolette, S.R., van der Kooy, D. Blockade of mesolimbic dopamine transmission dramatically increases sensitivity to the rewarding effects of nicotine in the ventral tegmental area. *Mol. Psychiatry.* 8:50, 2003.

81. Barrot, M., Olivier, J.D., Perrotti, L.I. et al. CREB activity in the nucleus accumbens shell controls gating of behavioral responses to emotional stimuli. *Proc. Natl. Acad. Sci. U.S.A.* 99:11435, 2002.

82. Benowitz, N.L., Porchet, H., Jacob, P., III. Nicotine dependence and tolerance in man: pharmacokinetic and pharmacodynamic investigations. *Prog. Brain Res.* 79:279, 1989.

83. Grady, S., Marks, M.J., Wonnacott, S., Collins, A.C. Characterization of nicotinic receptor-mediated [3H]dopamine release from synaptosomes prepared from mouse striatum. *J. Neurochem.* 59:848, 1992.

84. Russell, M.A. Subjective and behavioural effects of nicotine in humans: some sources of individual variation. *Prog. Brain Res.* 79:289, 1989.

85. Mitsis, E.M., van Dyck, C.H., Krantzler, E. et al. Prolonged occupancy of nicotinic acetylcholine receptors by nicotine in human brain: a preliminary study. Paper presented at the Annual Meeting of the Society for Research on Nicotine and Tobacco, Orlando, FL, 2006.

86. Rice, M.E., Cragg, S.J. Nicotine amplifies reward-related dopamine signals in striatum. *Nat. Neurosci.* 7:583, 2004.

87. Zhang, H., Sulzer, D. Frequency-dependent modulation of dopamine release by nicotine. *Nat. Neurosci.* 7:581, 2004.

88. Schultz, W. Getting formal with dopamine and reward. *Neuron.* 36:241, 2002.

89. Perkins, K.A., Gerlach, D., Vender, J., Grobe, J., Meeker, J., Hutchison, S. Sex differences in the subjective and reinforcing effects of visual and olfactory cigarette smoke stimuli. *Nicotine Tob. Res.* 3:141, 2001.

90. Rose, J.E., Behm, F.M. Extinguishing the rewarding value of smoke cues: pharmacological and behavioral treatments. *Nicotine Tob. Res.* 6:523, 2004.

91. Brody, A.L., Mandelkern, M.A., London, E.D. et al. Brain metabolic changes during cigarette craving. *Arch. Gen. Psychiatry.* 59:1162, 2002.

92. Due, D.L., Huettel, S.A., Hall, W.G., Rubin, D.C. Activation in mesolimbic and visuospatial neural circuits elicited by smoking cues: evidence from functional magnetic resonance imaging. *Am. J. Psychiatry.* 159:954, 2002.

93. Caggiula, A.R., Donny, E.C., Chaudhri, N., Perkins, K.A., Evans-Martin, F.F., Sved, A.F. Importance of nonpharmacological factors in nicotine self-administration. *Physiol. Behav.* 77:683, 2002.

94. Caggiula, A.R., Donny, E.C., White, A.R. et al. Cue dependency of nicotine self-administration and smoking. *Pharmacol. Biochem. Behav.* 70:515, 2001.

95. Caggiula, A.R., Donny, E.C., White, A.R. et al. Environmental stimuli promote the acquisition of nicotine self-administration in rats. *Psychopharmacology* (Berlin). 163:230, 2002.

96. Cohen, C., Perrault, G., Griebel, G., Soubrie, P. Nicotine-associated cues maintain nicotine-seeking behavior in rats several weeks after nicotine withdrawal: reversal by the cannabinoid (CB1) receptor antagonist, rimonabant (SR141716). *Neuropsychopharmacology.* 30:145, 2005.

97. Lesage, M.G., Burroughs, D., Dufek, M., Keyler, D.E., Pentel, P.R. Reinstatement of nicotine self-administration in rats by presentation of nicotine-paired stimuli, but not nicotine priming. *Pharmacol. Biochem. Behav.* 79:507, 2004.

98. Schroeder, B.E., Binzak, J.M., Kelley, A.E. A common profile of prefrontal cortical activation fol-
     lowing exposure to nicotine- or chocolate-associated contextual cues. *Neuroscience.* 105:535, 2001.
99. Kelley, A.E. Memory and addiction: shared neural circuitry and molecular mechanisms. *Neuron.*
     44:161, 2004.
100. Paterson, N.E., Markou, A. The metabotropic glutamate receptor 5 antagonist MPEP decreased break
     points for nicotine, cocaine and food in rats. *Psychopharmacology* (Berlin). 179:255, 2005.
101. Paterson, N.E., Semenova, S., Gasparini, F., Markou, A. The mGluR5 antagonist MPEP decreased
     nicotine self-administration in rats and mice. *Psychopharmacology* (Berlin). 167:257, 2003.
102. Bespalov, A.Y., Dravolina, O.A., Sukhanov, I. et al. Metabotropic glutamate receptor (mGluR5)
     antagonist MPEP attenuated cue- and schedule-induced reinstatement of nicotine self-administration
     behavior in rats. *Neuropharmacology.* 49(Suppl.):167, 2005.
103. Le Foll, B., Diaz, J., Sokoloff, P. Increased dopamine D3 receptor expression accompanying behavioral
     sensitization to nicotine in rats. *Synapse.* 47:176, 2003.
104. Le Foll, B., Schwartz, J.C., Sokoloff, P. Disruption of nicotine conditioning by dopamine D(3) receptor
     ligands. *Mol. Psychiatry.* 8:225, 2003.
105. Le Foll, B., Sokoloff, P., Stark, H., Goldberg, S.R. Dopamine D3 receptor ligands block nicotine-
     induced conditioned place preferences through a mechanism that does not involve discriminative-
     stimulus or antidepressant-like effects. *Neuropsychopharmacology.* 30:720, 2005.
106. Brunzell, D.H., Chang, J.R., Schneider, B., Olausson, P., Taylor, J.R., Picciotto, M.R. β2-Subunit-
     containing nicotinic acetylcholine receptors are involved in nicotine-induced increases in conditioned
     reinforcement but not progressive ratio responding for food in C57BL/6 mice. *Psychopharmacology*
     (Berlin). 184:328, 2006.
107. Olausson, P., Jentsch, J.D., Taylor, J.R. Repeated nicotine exposure enhances reward-related learning
     in the rat. *Neuropsychopharmacology.* 28:1264, 2003.
108. Olausson, P., Jentsch, J.D., Taylor, J.R. Repeated nicotine exposure enhances responding with condi-
     tioned reinforcement. *Psychopharmacology* (Berlin). 173:98, 2004.
109. Olausson, P., Jentsch, J.D., Taylor, J.R. Nicotine enhances responding with conditioned reinforcement.
     Psychopharmacology (Berlin). 171:173, 2004.
110. Palmatier, M.I., Peterson, J.L., Wilkinson, J.L., Bevins, R.A. Nicotine serves as a feature-positive
     modulator of Pavlovian appetitive conditioning in rats. *Behav. Pharmacol.* 15:183, 2004.
111. Le Foll, B., Goldberg, S.R. Cannabinoid CB1 receptor antagonists as promising new medications for
     drug dependence. *J. Pharmacol. Exp. Ther.* 312:875, 2005.
112. Silva, A.J., Kogan, J.H., Frankland, P.W., Kida, S. CREB and memory. *Annu. Rev. Neurosci.* 21:127,
     1998.
113. Sweatt, J.D. Mitogen-activated protein kinases in synaptic plasticity and memory. *Curr. Opin. Neu-
     robiol.* 14:311, 2004.
114. Brunzell, D.H., Russell, D.S., Picciotto, M.R. *In vivo* nicotine treatment regulates mesocorticolimbic
     CREB and ERK signaling in C57Bl/6J mice. *J. Neurochem.* 84:1431, 2003.
115. Pandey, S.C., Roy, A., Xu, T., Mittal, N. Effects of protracted nicotine exposure and withdrawal on
     the expression and phosphorylation of the CREB gene transcription factor in rat brain. *J. Neurochem.*
     77:943, 2001.
116. Valjent, E., Pages, C., Herve, D., Girault, J.A., Caboche, J. Addictive and non-addictive drugs induce
     distinct and specific patterns of ERK activation in mouse brain. *Eur. J. Neurosci.* 19:1826, 2004.
117. Walters, C.L., Cleck, J.N., Kuo, Y.C., Blendy, J.A. Mu-opioid receptor and CREB activation are
     required for nicotine reward. *Neuron.* 46:933, 2005.
118. Dineley, K.T., Westerman, M., Bui, D., Bell, K., Ashe, K.H., Sweatt, J.D. Beta-amyloid activates the
     mitogen-activated protein kinase cascade via hippocampal alpha7 nicotinic acetylcholine receptors:
     in vitro and *in vivo* mechanisms related to Alzheimer's disease. *J. Neurosci.* 21:4125, 2001.
119. Chang, K.T., Berg, D.K. Voltage-gated channels block nicotinic regulation of CREB phosphorylation
     and gene expression in neurons. *Neuron.* 32:855, 2001.
120. Nakayama, H., Numakawa, T., Ikeuchi, T., Hatanaka, H. Nicotine-induced phosphorylation of extra-
     cellular signal-regulated protein kinase and CREB in PC12h cells. *J. Neurochem.* 79:489, 2001.
121. Griffiths, J., Marley, P.D. Ca^{2+}-dependent activation of tyrosine hydroxylase involves MEK1. *Neu-
     roreport.* 12:2679, 2001.

122. Haycock, J.W. Multiple signaling pathways in bovine chromaffin cells regulate tyrosine hydroxylase phosphorylation at Ser19, Ser31, and Ser40. *Neurochem. Res.* 18:15, 1993.

123. Sparks, J.A., Pauly, J.R. Effects of continuous oral nicotine administration on brain nicotinic receptors and responsiveness to nicotine in C57Bl/6 mice. *Psychopharmacology* (Berlin). 141:145, 1999.

124. Pluzarev, O., Pandey, S.C. Modulation of CREB expression and phosphorylation in the rat nucleus accumbens during nicotine exposure and withdrawal. *J. Neurosci. Res.* 77:884, 2004.

125. Markou, A., Kenny, P.J. Neuroadaptations to chronic exposure to drugs of abuse: relevance to depressive symptomatology seen across psychiatric diagnostic categories. *Neurotox. Res.* 4:297, 2002.

126. Newton, S.S., Thome, J., Wallace, T.L. et al. Inhibition of cAMP response element-binding protein or dynorphin in the nucleus accumbens produces an antidepressant-like effect. *J. Neurosci.* 22:10883, 2002.

127. Kuehn, B.M. FDA speeds smoking cessation drug review. *JAMA* 8:295(6), 614, 2006.

128. O'Malley, S.S., Krishnan-Sarin, S., Farren, C., Sinha, R., Kreek, M.J. Naltrexone decreases craving and alcohol self-administration in alcohol-dependent subjects and activates the hypothalamo-pituitary-adrenocortical axis. *Psychopharmacology* (Berlin). 160:19, 2002.

129. Krishnan-Sarin, S., Meandzija, B., O'Malley, S. Naltrexone and nicotine patch smoking cessation: a preliminary study. *Nicotine Tob. Res.* 5:851, 2003.

130. Krishnan-Sarin, S., Rosen, M.I., O'Malley, S.S. Naloxone challenge in smokers. Preliminary evidence of an opioid component in nicotine dependence. *Arch. Gen. Psychiatry.* 56:663, 1999.

131. Rukstalis, M., Jepson, C., Strasser, A. et al. Naltrexone reduces the relative reinforcing value of nicotine in a cigarette smoking choice paradigm. *Psychopharmacology* (Berlin). 180:41, 2005.

132. Lesage, M.G., Keyler, D.E., Hieda, Y. et al. Effects of a nicotine conjugate vaccine on the acquisition and maintenance of nicotine self-administration in rats. *Psychopharmacology* (Berlin). 1, 2005.

CHAPTER 3

Neurochemical Substrates of Habitual Tobacco Smoking

Irina Esterlis, Ph.D., Suchitra Krishnan-Sarin, Ph.D., and Julie K. Staley, Ph.D.
Department of Psychiatry, Yale University School of Medicine, New Haven, Connecticut and VA Connecticut Healthcare System, West Haven, Connecticut

CONTENTS

Tobacco is the most widely abused substance in our society today. Not only are cigarettes highly addictive and the source of a multitude of social, economic, and medical consequences, but also their abuse is most prevalent among psychiatric populations, including persons afflicted with schizophrenia, bipolar, major depressive, anxiety, and substance abuse disorders. Cigarette smoking kills more Americans than accidents, alcoholism, fires, illegal drugs, AIDS, murder, and suicide combined, and is responsible for approximately 400,000 premature deaths per year in the U.S. and 4.83 million premature deaths per year worldwide.[1] The medical, social, and economic consequences of cigarette smoking cost the U.S. society approximately $100 billion annually.[2] Despite the overwhelming evidence of the medical risks associated with cigarette smoking, about 20% of the U.S. population continues to smoke. These devastating costs to society underscore the need for research into the neurochemical mechanisms underlying the development and maintenance of the addiction to cigarette smoking. By understanding the neurochemical substrates promoting the addiction to cigarettes, better treatments for this destructive and costly brain disorder may be developed.

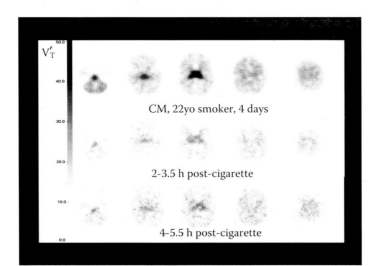

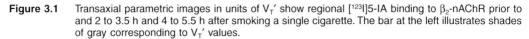

Figure 3.1   Transaxial parametric images in units of V_T' show regional [^{123}I]5-IA binding to β_2-nAChR prior to and 2 to 3.5 h and 4 to 5.5 h after smoking a single cigarette. The bar at the left illustrates shades of gray corresponding to V_T' values.

3.1 CHOLINERGIC ADAPTATIONS IN SMOKERS

The nicotinic acetylcholine receptor (nAChR) is the initial site of action of nicotine. With the advent of *in vivo* imaging methods, such as single photon emission computed tomography (SPECT), the amount of nicotine occupying nAChR in brain after smoking a cigarette may be measured. The occupancy of nAChR containing the β_2-subunit by nicotine after smoking one and two cigarettes has recently been determined using the nicotinic agonist radioligand [^{123}I]5-IA-85380 and SPECT. Occupancy of β_2-nAChR after smoking one cigarette ranged from 34 to 62%, while, after two cigarettes, the range was from 35 to 56%, both in a region-dependent manner. Interestingly, nicotine continually occupied β_2-nAChR 1.8 to 6 h after smoking a cigarette even in the presence of continued radiotracer infusion (see Figure 3.1). The long-lasting occupancy of the β_2-nAChR by nicotine raises important questions about the frequency of cigarette smoking. Specifically, why do smokers smoke cigarettes every 1 to 2 h if the receptor remains occupied by nicotine, a pharmacologically active metabolite or endogenous acetylcholine for up to 6 h after smoking? One hypothesis is that the long-lasting occupancy may render this subset of receptors inactive, thus promoting the upregulation of receptors and agonist-binding sites as has been noted in post-mortem brain and peripheral lymphocytes from smokers. [^{3}H]nicotine binding is higher in peripheral blood cells of smokers vs. nonsmokers and, interestingly, correlates with the number of cigarettes smoked per day.[3] [^{3}H]nicotine binding is higher in the gyrus rectus (Brodman area 11), hippocampus, thalamus, midbrain,[4,5] striatum, entorhinal cortex, and cerebellum,[6] and [^{3}H]epibatidine binding is higher in prefrontal and temporal cortex and hippocampus[7] in post-mortem brain from human smokers. Studies in animals treated chronically with nicotine have demonstrated that the upregulation in nAChR is due solely to the effects of nicotine.[8,9]

The mechanism of the upregulation in nicotine binding sites is not well understood. It has been established that, in contrast to the classic pharmacological dogma, which states that a desensitized receptor has lower affinity for agonists, the desensitized inactivated nAChR[10] exhibits higher binding affinity for agonists compared to the closed "resting" state and the open "activated" state.[11] When an endogenous agonist (e.g., acetylcholine) or an exogenous agonist (e.g., nicotine) binds to β_2-nAChR in its closed "resting" state, the channel undergoes a conformational change to the "open" state where ions influx. Subsequently, the ion channel undergoes a second conformational change to the "desensitized" state that corresponds to the closing of the channel. With prolonged desensitization, the β_2-

nAChR enters an "inactivated" state.[12] The state of nAChR receptor is determined by agonist concentration, and the time course of agonist administration.[11] It is well established that the agonist-induced conformational change of the nAChR to the desensitized state occurs rapidly (i.e., within milliseconds).[13] However, in animals the nAChR upregulation has been mapped at time points only as early as 7 to 24 h post-nicotine.[14,15] Thus, while it is known that the desensitized state of the receptor exhibits higher affinity for agonist that may give an appearance of an acute increase in β_2-nAChR, it appears that the increase in agonist binding to nAChR occurs only after prolonged inactivation of the receptor.[16] In contrast, a recent study demonstrated that increased high-affinity nicotine binding was paralleled by a twofold increase in acetylcholine-evoked currents that were less sensitive to desensitization. The differential reports relating function to increased binding may be due to different subunit combinations that all demonstrate increased agonist binding but differ in the effects on function. In keeping, Lindstrom and colleagues[17] recently demonstrated that doses of nicotine that activate $\alpha_3\beta_2$ block the channel, whereas the nicotine dose that maximally activates the $\alpha_4\beta_2$ combination does not block the channel. Increased nicotinic agonist binding is not associated with changes in β_2-nAChR mRNA,[10,18,19] and the role of protein synthesis is not clear (e.g., no effect).[20,21] There is increasing evidence that the upregulation results from a combination of increased receptor expression on the cell surface[22] and decreased receptor turnover,[10,23] and that this change is associated with persistent functional inactivation that occurs via distinct post-translational mechanisms and at rates and magnitudes that are nAChR-subtype specific.[12] Moreover, the magnitude of the effect of nicotine to upregulate nAChR agonist binding sites may be genetically determined. Feng and colleagues,[24,25] who studied the gene expression of α_4 and β_2 subunits in 901 male siblings from 222 families, noted a strong association between the severity of nicotine dependence and haplotypes of the α_4 and β_2 subunits in men. While more studies are needed, these preliminary findings suggest that the propensity to develop nicotine dependence may be genetically determined, which may explain why some smokers are able to smoke "casually" while others develop severe dependence.

Another important question relates to how long the receptor upregulation lasts. In rodents, [³H]nicotine binding is elevated in brain for up to 3 days, and normalizes to baseline values within 7 days of nicotine withdrawal,[26] whereas in living human tobacco smokers abstinent for 4 to 9 days, nAChR measured using [¹²³I]5-IA-85380 is elevated compared to age-matched never smokers suggesting that the receptor has not yet normalized at 1 week of abstinence.[27] In post-mortem human brain, high-affinity nicotine binding in ex-smokers (>2 months) is similar to that of the nonsmokers, suggesting that the receptor normalized within a 2-month period of time.[4–6] In a preliminary sample of living human smokers, this time frame is similar.[28] Thus, while the time period necessary for normalization is still unclear, it is apparent that the time frame for normalization in humans is longer than that noted in rodents.

An important consideration for measuring the nicotine binding site on nAChR in living humans is the time interval since the last cigarette required for residual nicotine to clear from brain so that it does not interfere with binding of the radioligand. Studies in nonhuman primates indicated that approximately 7 days would be required for nicotine to clear.[29] Thus, measurements of nAChR levels in humans should be obtained within a time interval in which the upregulation is still evident, but yet, sufficient time for nicotine to clear has been achieved. An important note is that plasma nicotine levels, which have a half-life of approximately 2 h, are a poor indicator of clearance of nicotine or pharmacologically active metabolites from brain. In our studies of both nonhuman primates chronically administered nicotine and human tobacco smokers, we have found urine cotinine levels to be a reliable predictor of nicotine clearance from brain.

3.2 DOPAMINERGIC ADAPTATIONS IN SMOKERS

Alterations in dopamine (DA) levels are associated with the rewarding effects of abused substances including cigarettes. Specifically, the mesolimbic DA pathway, which originates in the ventral tegmental area (VTA) and projects to nucleus accumbens, is believed to be the primary reward pathway in the brain.[30] nAChRs containing the alpha7 subunit (α_7-nAChR) are abundant in the VTA. Stimulation of these receptors by nicotine or by endogenous acetylcholine, whose release was induced by nicotine or other components of cigarette smoke, produces an increase in glutamate concentrations which in turn stimulate N-methyl-D-aspartate receptors (NMDARs) on DA-containing neurons in the VTA, facilitating DA release and enhancing dopaminergic function in this critical brain reward area.[31] α_7-nAChRs are localized on glutamatergic terminals along with β_2-nAChRs on gamma-aminobutyric acid (GABA) nerve terminals postsynaptic to DA neurons within the VTA. Additionally, β_2-nAChRs are localized to DA cell bodies within the VTA. Nicotine actions on each of these strategically localized nAChRs are likely responsible for the interactions of nicotine with the DA reward pathway that mediates the development and maintenance of habitual tobacco smoking.[32]

In human tobacco smokers, synaptic DA levels increase in response to smoking a single cigarette.[33] [^{11}C]raclopride binding to DA D_2 receptors is sensitive to endogenous DA levels and, thus, DA release may be determined by measuring the change in [^{11}C]raclopride binding to D_2 receptors after smoking a cigarette. After smoking a cigarette, [^{11}C]raclopride binding was reduced by 25.9 to 36.6% compared to the 0 to 13.6% decrease observed in smokers who did not smoke a cigarette, demonstrating that smoking a cigarette causes significant DA release in the striatal reward areas. Moreover, DA levels positively correlated with craving in left ventral caudate/nucleus accumbens ($r = 0.49$, $p = 0.04$) and in putamen ($r = 0.65$, $p = 0.004$), suggesting that the larger DA release provided a greater relief from craving. In a similar study design in nonhuman primates, nicotine caused a 5 to 6% reduction in [^{11}C]raclopride binding to D_2 receptors after a nicotine infusion in nonhuman primates.[34] The lower amount of DA release from nicotine challenge compared to a smoking challenge is not surprising and suggests that other chemical(s) in tobacco smoke may be contributing to the reinforcing properties of smoking by enhancing nicotine-induced DA release. Fowler and colleagues[35] have elegantly demonstrated that monoamine oxidase B (MAO-B, the primary catabolic enzyme for DA in the brain) is lower throughout the brain of living smokers (basal ganglia, thalamus, cerebellum, cingulate gyrus, and frontal cortex). Similar decreases have been noted in post-mortem brain of tobacco smokers where lower [^{3}H]azabemide binding to MAO-B in amygdala was observed.[36] Because nicotine does not inhibit MAO-B,[37] it appears that the lower MAO-B levels in smoker's brain is due to chronic inhibition by other components of tobacco smoke. For example, tobacco smoke contains harmala alkaloids, including harmon and norharmon, that are potent monoamine oxidase inhibitors (MAOIs),[38] and Villegier[39] observed behavioral sensitization induced by repeated injections of nicotine in rats is short-lasting, but was prolonged upon the co-injection of a MAOI. These findings imply that behavioral effects of nicotine are transient and insufficient to induce long-term behavioral sensitization in the absence of MAOIs, suggesting that MAOIs contribute to the addictive properties of tobacco smoking. Collectively, these studies lead to the conclusion that the rewarding properties of tobacco smoking are mediated by the combined effects of nicotine and harmala alkaloids on DA release.

In addition to the acute effects of tobacco smoking on dopaminergic function in the striatal reward areas, there is significant evidence suggesting that the smoker's brain is in a chronic hyperdopaminergic state. Uptake of L-dopa (the precursor to DA) is higher in smokers vs. nonsmokers, suggesting DA biosynthesis is accelerated in smokers. In keeping, the striatal homovanillic acid (HVA)/DA ratio is lower in post-mortem brain of smokers compared to non-smokers due to higher DA levels (as opposed to lower HVA).[6] It is interesting to note that tyrosine hydroxylase (TH), the rate-limiting enzyme for DA biosynthesis, has been associated with vulnerability to develop nicotine dependence. Specifically, smokers that have the K4 allele of the TH enzyme are about 85% less likely to smoke in a dependent manner.[40] On the other hand, those carrying the K1 allele or three single nucleotide polymorphisms (SNPs) at the TH locus were not protected from developing nicotine dependence upon smoking. While the relationship of the K4 allele to TH expression and function is unclear, one may speculate that individuals with the K4 allele may synthesize DA at a slower rate, resulting in a smaller increase in DA in response to smoking and in turn decreasing reward salience and vulnerability to developing nicotine dependence.

The DA transporter has been suggested to be a critical dopaminergic substrate for habitual tobacco smoking because of its innate function — to regulate intrasynaptic DA availability and dopaminergic neurotransmission — and also because it has been genetically linked to nicotine dependence and age of smoking onset.[41] However, studies of DA transporter availability in post-mortem human brain from elderly tobacco smokers[6] and also in a younger population of living smokers did not detect a significant difference in striatal DA transporter availability. Moreover, there appears to be no significant relationship between DA transporter availability and smoking behavior.[42] Interestingly, the vesicular monoamine transporter (VMAT$_2$), an intra-neuronal carrier among all monoaminergic systems, is decreased in platelets of habitual smokers vs. nonsmokers.[43]

With regards to DA receptors, postsynaptic DA D$_1$ receptors are decreased in living smokers as evidenced by lower [^{11}C]SCH 23390 binding in smokers compared to nonsmokers in the striatum, and most specifically in the nucleus accumbens.[44] Since [^{11}C]SCH 23390 binding to striatal D$_1$ receptors is insensitive to acute changes in extracellular DA concentration *in vivo*, it is likely that the decrease truly reflects altered D$_1$ receptor availability in smokers. Of note, the only study that has assessed D$_1$ receptors in post-mortem human brain of smokers compared to nonsmokers did not note any differences in D$_1$ receptor number.[6] The discrepancy between the post-mortem and *in vivo* study in living smokers may be due to age differences in the subject populations studied, or may reflect confounds associated with studies in post-mortem tissue including length of storage time and post-mortem interval. Lower D$_{1-like}$ receptor availability as observed in living smokers is a logical compensatory adaptive response to prolonged repeated perturbations in elevated synaptic DA levels. Alternatively, lower D$_{1-like}$ receptor availability may be genetically determined. Currently, there is no evidence for a genetic relationship between the D$_1$ receptor genotype and smoking behaviors; however, preliminary evidence suggests that the T allele of the closely related D$_5$ receptor is protective against smoking initiation.[45] Because there are no drugs available that pharmacologically distinguish between D$_1$ and D$_5$ receptors, the relationship between D$_{1/5}$ genotypes and receptor availability is unclear. However, it may be hypothesized that smokers smoke in effort to enhance dopaminergic signaling of an innately lower dopaminergic state that would make them more vulnerable to developing an addiction to tobacco smoking.

D$_2$ receptor function also appears to be aberrant in smokers as demonstrated by reduced growth hormone response to apomorphine challenge compared with nonsmokers. Interestingly, there was no difference between response to apomorphine during *ad libitum* smoking and 12 h of abstinence. These findings suggest that regardless of whether or not nicotine is on board, D$_2$ receptor sensitivity to DA agonists is reduced in smokers.[46] Importantly, D$_2$ receptor sensitivity to apomorphine was inversely correlated with cotinine serum levels and severity of nicotine dependence as measured by the Fagerstrom Tolerance Nicotine Dependence questionnaire (FTND). Reduced sensitivity of D$_2$ receptors to agonist stimulation likely reflects uncoupling of G proteins from the D$_2$ receptors, which would decrease the sensitivity to agonist stimulation but would not demonstrate a difference in D$_2$ receptor numbers measured using a radiolabeled antagonist.[6] Further support for the D$_2$ receptor as a neurochemical substrate of smoking is provided by evidence demonstrating that smokers carrying the A1 allele of the D$_2$ receptor have reduced P300 amplitude compared to nonsmokers[47] and carriers of the rare B1 allele for the D$_2$ receptor gene are more likely to be ever smokers.[48] In addition, the D$_4$ VNTR polymorphism moderates reactivity to smoking cues. Specifically, carriers of the DRD$_4$ L polymorphism demonstrate greater craving, attention, and arousal in response to smoking cues.[49] In contrast, individuals carrying the D$_4$ S allele do not demonstrate reactivity to smoking cues. Collectively, these studies support the D$_2$-like receptors as critical substrate for vulnerability to tobacco smoking.

3.3 GABAergic ADAPTATIONS IN SMOKERS

GABA is a major neurotransmitter in the mammalian brain and controls neuronal excitability. It has been implicated in the addictive and withdrawal processes of nicotine dependence. Nicotine stimulates GABA release via modulation of nAChR on GABAergic neurons, which could lead to a decrease in inhibitory tone from GABAergic stimulation of $GABA_B$ autoreceptors.[50] Alternatively, nicotine-induced alterations in the levels of neurosteroids that regulate $GABA_A$ receptors could also potentially lead to altered levels of GABA.[51] Changes in GABA might also occur through nicotine's actions at the nAChR, which increases GABA release. Nicotine-induced GABA release is blocked by mecamylamine and dihydro-β-erythroidine and the effect is lost in the $β_2$ knockout mice, suggesting that $β_2$-nAChR mediates GABA release. The $α_7$ nAChR antagonist, alpha bungarotoxin, did not alter release.[52,53] Nicotine-induced GABA release has been demonstrated in the thalamus, hippocampus, and throughout the cerebral cortex.[54–60] The differential effects of nicotine on GABA release may be due to regional differences in nAChR subunit combinations in different regions.

Cortical GABA is also dysregulated in disorders associated with affective instability,[61] including premenstrual dysphoric disorder. Since nicotine modulates GABA function, Epperson and colleagues[62] suggest it is possible that nicotine modulates mood.[62] In a magnetic resonance spectroscopy (MRS) study of men and women smokers abstinent for 48 h, cortical GABA levels were decreased in women smokers imaged during the follicular phase (when hormone levels are similar to those in men) as compared to men. Furthermore, there were no differences in GABA levels between men smokers and nonsmokers but there was a drastic decrease in GABA levels in women smokers compared to women nonsmokers during the follicular phase of the menstrual cycle. These findings suggested that the phasic differences in cortical GABA levels evident in women nonsmokers are suppressed in women smokers. Since menstrual cycle phase was confirmed by serum estradiol and progesterone levels, changes in GABA levels cannot be attributed to lack of hormonal cyclicity.

The acute or chronic regulatory effects of nicotine treatment or tobacco smoking on cortical $GABA_A$-benzodiazepine receptors ($GABA_A$–BZR) are poorly studied. To date, only one animal study has examined effects of chronic nicotine and has demonstrated increased $GABA_A$–BZ receptors.[63] In addition to nicotine, tobacco smoke also contains the β-carbolines harman and norharman,[64] which are well known as MAOIs,[65,66] but may also be inverse agonists at the $GABA_A$–BZ receptor.[67] Currently, it is not clear what the combined effects of nicotine and β-carbolines are on $GABA_A$–BZR.

There is some evidence for a role for $GABA_B$ receptors as a neurochemical substrate of tobacco smoking. $GABA_{B1}$ receptors appear to be regulated by nicotine. Li[50] demonstrated a significant reduction of $GABA_B$ receptor mRNA in the hippocampus in rats chronically treated with nicotine. $GABA_B$ receptors primarily function to modulate release of neurotransmitters including GABA, glutamate, acetylcholine, noradrenalin, and serotonin that in the hippocampus are important for cognitive processes including attention and memory. Thus, nicotine-induced alterations in $GABA_B$ receptor expression in the hippocampus, a widely accepted site for learning and memory in both humans and animals, may be implicated in cognitive properties of tobacco smoke on cognition. A genetic linkage between $GABA_{B2}$ and nicotine dependence has been demonstrated in African American and European Americans.[68]

3.4 OPIOIDERGIC ADAPTATIONS IN SMOKERS

The endogenous opioid system is believed to be the primary common pathway for all drugs of abuse. However, the role of the opioid system in habitual tobacco smoking has only recently become of interest. Using the short-acting mu-opioid antagonist naloxone, some studies have found decreases in smoking behavior in short-term laboratory paradigms[69,70] while others have reported no effect of naloxone on smoking behavior.[71] The long-acting mu-opioid antagonist naltrexone has been studied more extensively and has been shown to reduce smoking behavior and craving for cigarettes.[72,73] When used in combination with the nicotine patch, naltrexone has been shown to reduce smoking behavior and tobacco craving[74] and craving in response to cues,[75] as well as to block some effects of nicotine.[76] Evidence from clinical trials of naltrexone is equivocal with both positive[74,77] and negative findings.[78,79] Preclinical evidence suggests that the effects of naltrexone on cigarette smoking may be mediated by its ability to differentially alter expression and function of the α_7 and $\alpha_4\beta_2$ nAChRs in the central nervous system.[80]

Nicotine administration causes release of the endogenous opioid peptide beta-endorphin.[81] Preclinical evidence suggests that the mu-opioid receptors are involved in nicotine reward.[82] Nicotine induced sufficient beta-endorphin to displace binding of the mu-opioid receptor agonist ^{11}C-carfentanil in brain in recently abstinent male smokers. Importantly, Scott and colleagues[83] used [11C]carfentanil and positron emission tomography (PET) to demonstrate that nicotine, but not denicotinized cigarettes, induced endogenous opioid release in the thalamus and amygdala but reduced release in the anterior cingulate suggesting that nicotine was the sole chemical in tobacco smoke responsible for activation of the opioid reward pathway in the thalamus, amygdala, and cingulate. This suggests opioid changes in the cingulate mediate craving for nicotine while the opioidergic changes in the thalamus and amygdala may mediate the feelings of "satisfaction" after smoking a cigarette.

Lerman and colleagues[84] proposed that OPRM1 (mu-opioid receptor gene) may be responsible for efficacy of the type of nicotine replacement therapy and examined the relationship of the OPRM1 in relation to response to different nicotine replacement therapies. Smokers carrying the less common OPRM1 Asp40 variant were significantly more likely than those homozygous for the wild-type Asn40 variant to be abstinent at the end of nicotine replacement phase, with the effect being significant for the transdermal nicotine vs. nicotine nasal spray therapies. Furthermore, individuals with Asp40 variant treated by transdermal nicotine exhibited a significantly higher rate of recovery from short smoking lapses than those with Asn40 variant and significantly less negative side effects of smoking cessation (e.g., weight gain and withdrawal symptoms). These findings suggest that nicotine replacement therapies will be more efficacious in carriers of the Asp40 variant for the opioid receptor.

3.5 SEROTONERGIC ADAPTATIONS IN SMOKERS

Serotonin (5-HT) regulates many bodily functions, including appetite[85] and sleep (i.e., modulation of REM latency),[86] and may be involved in initiation and maintenance of tobacco smoking. Drugs that enhance 5-HT levels facilitate smoking cessation in highly dependent smokers.[87–89] In turn, nicotine has been shown to elevate 5-HT levels by stimulating 5-HT release through binding to the nAChR[90] and inhibiting 5-HT reuptake.[90,91] 5-HT levels are further enhanced in the smoker's brain as a consequence of decreased monoamine oxidase-A (MAO-A; the neuronal enzyme that serves to degrade 5-HT) activity.[35] Similar decreases in MAO activity have been noted in platelets of smokers[92] and support reports of twofold higher platelet 5-HT levels in smokers as compared to nonsmokers.[93] In addition, nicotine has been shown to decrease platelet 5-HT release and inhibit 5-HT uptake.[94] Active smokers excrete approximately 30% more 5-HT and 5-hydroxyindoleacetic acid (5-HIAA) as compared to never smokers and former smokers.[95] In the central nervous system, nicotine and its primary metabolite cotinine both decrease 5-HT turnover in rat brain, which results in a net enhancement in 5-HT neurotransmission. The 5-HT transporter (5HTT) regulates magnitude and duration of serotonergic neurotransmission. Chronic exposure to nicotine is associated with reduction in 5HTT sites in the brain[96] and nicotine dependence (as assessed using FTND questionnaire) has been found to be inversely correlated with densities of platelet 5HTT.[97,98] In brain, diencephalon 5-HT transporter availability is not altered in living human tobacco smokers.[42] However, there is a trend for higher brain stem 5-HT transporter availability in smokers vs. nonsmokers (10% higher in smokers), which appears to be more evident in men than women smokers. The perturbations in 5-HT function induced by nicotine may in part contribute to the reinforcing properties of cigarette smoking. In keeping, smoking cessation is facilitated by enhancing 5-HT function by administration of the MAO-A inhibitor moclobemide in highly dependent smokers.[99]

5-HT$_{1A}$ receptor gene expression is higher in DG, C1, and C3 subfields of the hippocampus after 2 and 24 h nicotine administration in rodents,[100] suggesting that nicotine is capable of modulating 5-HT$_{1A}$ receptor expression in some cortical and limbic brain regions. Rasmussen and Szachura[101] examined the effects of 5-HT$_{1A}$ agonist 8-OH-DPAT on the single-unit activity of serotonergic neurons in anesthetized rats undergoing nicotine withdrawal. They demonstrated a significant increase in the DRN to the 5-HT$_{1A}$ agonist 8-OH-DPAT during nicotine withdrawal, which led to an enhanced startle response. They report an increase in sensitivity develops over time with significance at days 3 and 4, and dropping to baseline by day 7. This finding may suggest that pre- and post-synaptic 5-HT$_{1A}$ antagonist drugs may be useful in attenuating some of the symptoms of nicotine withdrawal, therefore contributing to smoking cessation in humans. 5-HT$_{2A}$ receptors play a role in schizophrenia and alcohol dependence, both of which are associated with high prevalence of smoking behavior. However, the specific role of this receptor in smoking has not been widely studied. One study showed an association between 5-HT$_{2A}$ and maintenance of smoking but not smoking initiation.[102]

3.6 IMPLICATIONS FOR SMOKING CESSATION TREATMENTS

Tobacco smoking is currently the most prevalent and deadly addiction. While there are numerous treatments currently available, there is a lot of room for improvement. Two types of pharmacological therapies have been approved by the U.S. Food and Drug Administration (FDA) — nicotine replacement therapies including gum, transdermal patch, lozenge, and inhaler, which deliver nicotine without the tar, and non-nicotine-based therapy such as bupropion hydrochloride (Zyban). There is significant between-subject variability in the efficacy of nicotine and non-nicotine-based pharmacotherapies, which could play a role in individual ability to quit and abstain from tobacco smoking. Factors such as genetic susceptibility, including family history, are currently being investigated in an effort to enhance the effectiveness of pharmacotherapies for smoking cessation. Because of the critical roles in drug reward, DA and opioid substrates are candidates for smoking cessation pharmacotherapies. Stimulation of D$_2$ receptors via bromocriptine decreases smoking, whereas D$_2$ receptor antagonism via haloperidol facilitates smoking. Zyban (bupropion), an atypical anti-depressant, has demonstrated efficacy for promoting long-term abstinence by reducing nicotine-related withdrawal symptoms,[103] negative affect,[104] and craving.[105] Zyban's mechanism of action for reducing smoking is believed to be inhibition of DA and norepinephrine reuptake, enhancement of norepinephrine and 5-HT neuronal activity, as well as noncompetitive inhibition of $\alpha_3\beta_2$, $\alpha_4\beta_2$, and α_7 nAChRs. However, Zyban is not equally effective in all smokers. For example, David and colleagues[106] demonstrated that individuals with *DRD2-Taq1 A2/A3* experience less craving upon smoking cessation, and reduced anxiety and impatience as compared to those with *DRD2-Taq1 A1/A2* or *A1/A1* who demonstrated no reduction in withdrawal symptoms.

Several other clinically available pharmacological agents have been tested for their potential to facilitate smoking cessation, although they are not approved by the FDA for this purpose. For example, tricyclic antidepressants, which inhibit reuptake of noradrenaline and 5-HT, promote smoking cessation in conjunction with behavioral treatment in some individuals.[107] However, these medications are limited because of their significant side effects. 5-HT-selective reuptake inhibitors (SSRIs) are believed to be a safer class of antidepressants but have not demonstrated effectiveness in smoking cessation.[108]

Cohen proposed that there may be a third way to treat nicotine dependence.[109] Since smokers still experience withdrawal symptoms with bupropion, and nicotine replacement therapies are not fully effective, Cohen and colleagues examined the effect of using a nicotinic receptor agonist in order to aid in smoking cessation. A novel nAChR ligand SSR591813 was employed due to its selective $\alpha_4\beta_2$ partial agonist activity. SSR591813 reduced the number of nicotine infusions on day 2 and 3 of treatment. Unlike mecamylamine, SSR591813 did not precipitate withdrawal signs in nicotine-exposed rats but prevented withdrawal signs precipitated by mecamylamine. Cohen and colleagues suggest these results imply $\alpha_4\beta_2$ involvement in the nicotine withdrawal syndrome. Since the SSR591813 may moderate nicotine withdrawal symptoms, which have been shown to cause enough distress to individuals that they relapse, it is important to continue investigation in its use for smoking cessation.

ACKNOWLEDGMENTS

This work was supported by R01DA015577 and Transdisciplinary Tobacco Research Center (P50AA15632).

REFERENCES

1. Ezzati, M., Lopez, A. Estimates of global mortality attributable to smoking in 2000. *Lancet.* 362:847, 2003.
2. NIDA. NIDA InfoFacts: Cigarettes and Other Nicotine Products; 2005.
3. Benhammou, K.L.M., Strook, M.A., Sullivan, B., Logel, J., Raschen, K., Gotti, C., Leonard, S. [³H]nicotine binding in peripheral blood cells of smokers is correlated with the number of cigarettes smoked per day. *Neuropharmacology.* 39:2818, 2000.
4. Benwell, M., Balfour, D., Anderson, J. Evidence that tobacco smoking increases the density of (–)-[³H]nicotine binding site in human brain. *J. Neurochem.* 50:1243, 1988.
5. Breese, C., Marks, M., Logel, J. et al. Effect of smoking history on [3H]nicotine binding in human postmortem brain. *J. Pharmacol. Exp. Ther.* 282:7, 1997.
6. Court, J., Lloyd, S., Thomas, N. et al. Dopamine and nicotinic receptor binding and the levels of dopamine and homovanillic acid in human brain related to tobacco use. *Neuroscience.* 87:63, 1998.
7. Perry, D., Dávila-García, M., Stockmeier, C., Kellar, K. Increased nicotinic receptors in brains from smokers: Membrane binding and autoradiography studies. *J. Pharmacol. Exp. Ther.* 289:1545, 1999.
8. Bhat, R., Turner, S., Selvaag, S., Marks, M., Collins, A. Regulation of brain nicotinic receptors by chronic agonist infusion. *J. Neurochem.* 56:1932, 1991.
9. Mochizuki, T., Villemagne, V., Scheffel, U. et al. Nicotine induced up-regulation of nicotinic receptors in CD-1 mice demonstrated with an *in vivo* radiotracer: gender differences. *Synapse.* 30:116, 1998.
10. Peng, X., Gerzanich, V., Anand, R., Whiting, P., Lindstrom, J. Nicotine-induced increase in neuronal nicotinic receptor results from a decrease in the rate of receptor turnover. *Mol. Pharmacol.* 46:523, 1994.
11. Dani, J., Heinemann, S. Molecular and cellular aspects of nicotine abuse. *Neuron.* 16:905, 1996.
12. Ke, L., Eisenhour, C., Bencherif, M., Lukas, R. Effects of chronic nicotine treatment on expression of diverse nicotinic acetylcholine receptor subtypes. I. Dose- and time-dependent effects of nicotine treatment. *J. Pharmacol. Exp. Ther.* 286:825, 1998.
13. Ochoa, E., Chattopadhyay, A., McNamee, M. Desensitization of the nicotinic acetylcholine receptor: Molecular mechanisms and effect of modulators. *Cell Mol. Neurobiol.* 9:141, 1989.
14. Marks, M., Stitzel, J., Collins, A. Time course study of the effects of chronic nicotine infusion on drug response and brain receptors. *J. Pharmacol. Exp. Ther.* 235:619, 1985.
15. Nasrallah, H., Coffman, J., Olson, S. Structural brain imaging findings in affective disorders: an overview. *J. Neuropsychiatr. Clin. Neurosci.* 1:21, 1989.
16. Wonnacott, S. The paradox of nicotinic acetylcholine receptor upregulation by nicotine. *Trends Pharmacol. Sci.* 11:216, 1990.

17. Rush, R., Kuryatov, A., Nelwon, M., Lindstrom, J. First and second transmembrane segments of α_3, α_4, β_2 and β_4 nicotinic acetylcholine receptor subunits influence the efficacy and potency of nicotine. *Mol. Pharmacol.* 61:1416, 2002.

18. Marks, M., Pauly, J., Gross, S. et al. Nicotine binding and nicotinic receptor subunit RNA after chronic nicotine treatment. *J. Neurosci.* 2765, 1992.

19. Zhang, X., Gong, Z.-H., Hellstrom-Lindahl, E., Nordberg, A. Regulation of $\alpha_4\beta_2$ nicotinic acetylcholine receptors in M10 cells following treatment with nicotinic agents. *Neuroreport.* 6:313, 1994.

20. Buisson, B., Bertrand, D. Chronic exposure to nicotine upregulates the human $\alpha_4\beta_2$ nicotinic acetylcholine receptor function. *J. Neurosci.* 21:1819, 2001.

21. Gopalakrishnan, M., Molinari, E., Sullivan, J. Regulation of human $\alpha_4\beta_2$ neuronal nicotinic acetylcholine receptors by cholinergic channel ligands and second messenger pathways. *Mol. Pharmacol.* 52:524, 1997.

22. Harkness, P., Millar, N. Changes in conformation and subcellular distribution of a4b2 nicotinic acetylcholine receptors revealed by chronic nicotine treatment and expression of subunit chimeras. *J. Neurosci.* 22:10172, 2002.

23. Wang, F., Nelson, M., Kuryatov, A. et al. Chronic nicotine treatment upregulates human $\alpha_3\beta_2$ but not $\alpha_3\beta_4$ acetylcholine receptors stably transfected in human embryonic kidney cells. *J. Biol. Chem.* 1998:28721, 1998.

24. Feng, Y.N.T., Xing, H., Xu, X., Chen, C., Peng, S., Wang, L., Xu, X. A common haplotype of the nicotine acetylcholine receptor alpha-4 subunit gene is associated with vulnerability to nicotine addiction in men. *Am. J. Hum. Genet.* 75:112, 2004.

25. Feng, Y., Niu, T., Xing, H. et al. A common haplotype of the nicotine acetylcholine receptor α_4 subunit gene is associated with vulnerability to nicotine addiction in men. *Am. J. Hum. Genet.* 75:112, 2004.

26. Pietila, K., Lahde, T., Attila, M., Ahtee, L., Nordberg, A. Regulation of nicotinic receptors in the brain of mice withdrawn from chronic oral nicotine treatment. *Naunyn-Schmiedeberg's Arch. Pharmacol.* 357:176, 1998.

27. Staley, J., Krishnan-Sarin, S., Cosgrove, K. et al. β_2* Nicotinic acetylcholine receptor availability in recently abstinent smokers. *J. Neurosci.* Accepted.

28. Cosgrove, K.P., Frohlich, E.B., Krantzler, E. et al. SPECT imaging of beta 2 nicotine acetylcholine receptors in tobacco smokers during acute and prolonged withdrawal. Paper presented at the Annual Meeting for the Society of Research on Nicotine and Tobacco, 2006.

29. Cosgrove, K., Ellis, S., Al-Tikriti, M. et al. Assessment of the effects of chronic nicotine on β_2-nicotinic acetylcholine receptors in nonhuman primate using [I-123]5-IA-85830 and SPECT. Paper presented at the 66th Annual Scientific Meeting of the College on Problems of Drug Dependence, 2004, San Juan, Puerto Rico.

30. Walters, C.L., Kuo, Y.C., Blendy, J.A. Differential distribution of CREB in the mesolimbic dopamine reward pathway. *J. Neurochem.* 87:1237, 2003.

31. Nomikos, G.G., Schilstrom, B., Hilderbrand, B.E., Panagis, G., Grenhoff, J., Svensson, T.H. Role of alpha7 nicotinic receptors in nicotine dependence and implications for psychiatric illness. *Behav. Brain Res.* 113:97, 2000.

32. Pich, E., Pagliusi, S., Tessari, M., Talabot-Ayer, D., Huijsduijnen, R.H.v., Chiamulera, C. Common neural substrates for the addictive properties of nicotine and cocaine. *Science.* 275:83, 1997.

33. Brody, A.L., Olmstead, R.E., London, E.D., Farahi, J., Meyer, J.H., Grossman, P., Lee, G.S., Huang, J., Hahn, E.L., Mandelkern, M.A. Smoking-induced ventral striatum dopamine release. *Am. J. Psychiatry.* 161:1211, 2004.

34. Marenco, S., Carson, R., Berman, K., Herscovitch, P., Weinberger, D. Nicotine-induced dopamine release in primates measured with [^{11}C]raclopride PET. *Neuropsychopharmacology.* 259, 2004.

35. Fowler, J., Volkow, N., Wang, G.-J. et al. Brain monoamine oxidase A inhibition in cigarette smokers. *Proc. Natl. Acad. Sci. U.S.A.* 93:14065, 1996.

36. Karolewicz, B., Klimek, V., Zhu, H., Szebeni, K., Nail, E., Stockmeier, C.A., Johnson, L., Ordway, G.A. Effects of depression, cigarette smoking, and age on monoamine oxidase B in amygdaloid nuclei. *Brain Res.* 1043:57, 2005.

37. Fowler, J., Volkow, N., Wang, G. et al. Neuropharmacological actions of cigarette smoke: brain monoamine oxidase B (MAO B) inhibition. *J. Addictive Dis.* 17:23, 1998.

38. Nelson, D., Herbet, A., Glowinski, J., Harmon, M. [3H]Harmaline as a specific ligand of MAO A. II. Measurement of the turnover rates of MAO A during ontogenesis in the rat brain. *J. Neurochem.* 32:1829, 1979.

39. Villegier, A.S., Blanc, G., Glowinski, J., Tassin, J.P. Transient behavioral sensitization to nicotine becomes long-lasting with monoamine oxidases inhibitors. *Pharmacol. Biochem. Behav.* 76:267, 2003.

40. Anney, R.J.L., Olsson, C.A., Lotfi-Miri, M., Patton, G.C., Williamson, R. Nicotine dependence in a prospective population-based study of adolescents: the protective role of a functional tyrosine hydroxylase polymorphism. *Pharmacogenetics.* 14:73, 2004.

41. Ling, D., Niu, T., Feng, Y., Xing, H., Xu, X. Association between polymorphism of the dopamine transporter gene and early smoking onset: an interaction risk on nicotine dependence. *J. Hum. Genet.* 49:35, 2004.

42. Staley, J., Krishnan-Sarin, S., Zoghbi, S. et al. Sex differences in [123I]beta-CIT SPECT measures of dopamine and serotonin transporter availability in healthy smokers and nonsmokers. *Synapse.* 41:275, 2001.

43. Schwartz, K., Weizman, A., Rehavi, M. Decreased platelet vesicular monoamine transporter density in habitual smokers. *Eur. Neuropsychopharmacol.* 15:235, 2005.

44. Dagher, A., Bleicher, C., Aston, J.A.D., Gunn, R.N., Clarke, P.B.S., Cumming, P. Reduced dopamine D1 receptor binding in the ventral striatum of cigarette smokers. *Synapse.* 42:48, 2001.

45. Sullivan, P.F., Neale, M.C., Silberman, M.A., Harris-Kerr, C., Myakishev, M., Wormley, B., Webb, B.T., Ma, Y., Kendler, K.S., Straub, R.E. An association study of DRD5 with smoking initiation and progression to nicotine dependence. *Am. J. Med. Genet.* 105:259, 2001.

46. Smolka, M.N., Budde, H., Karow, A.C., Schmidt, L.G. Neuroendocrinological and neuropsychological correlates of dopaminergic function in nicotine dependence. *Psychopharmacology.* 175:374, 2004.

47. Anokhin, A.P., Torodov, A.A., Madden, P.A.F., Grant, J.D., Heath, A.C. Brain event-related potentials, dopamine D2 receptor gene polymorphism, and smoking. *Genet. Epidemiol.* 17(Suppl. 1):S37, 1999.

48. Spitz, M.R., Shi, H., Yang, F. et al. Case-control study of the D2 dopamine receptor gene and smoking status in lung cancer patients [see comment]. *J. Natl. Cancer Inst.* 90:358, 1998.

49. Hutchison, K.E., LaChance, H., Niaura, R., Bryan, A., Smolen, A. The DRD4 VNTR polymorphism influences reactivity to smoking cues. *J. Abnormal Psychol.* 111:134, 2002.

50. Li, S., Park, M., Bahk, J., Kim, M. Chronic nicotine and smoking exposure decreases $GABA_{B1}$ receptor expression in the rat hippocampus. *Neurosci. Lett.* 334:135, 2002.

51. Porcu, P., Sogliano, C., Cinus, M., Purdy, R.H., Biggio, G., Concas, A. Nicotine induced changes in cerebrocortical neuroactive steroids and plasma corticosterone concentrations in the rat. *Pharmacol. Biochem. Behav.* 74:683, 2003.

52. Lena, C., Changeux, J.-P. Role of Ca^{2+} ions in nicotinic facilitation of GABA release in mouse thalamus. *J. Neurosci.* 17:576, 1997.

53. Lu, Y., Grady, S., Marks, M., Picciotto, M., Changeux, J.-P., Collins, A. Pharmacological characterization of nicotinic receptor stimulated GABA release from mouse brain synaptosomes. *J. Pharmacol. Exp. Ther.* 287:648, 1998.

54. Meshul, C.K., Kamel, D., Moore, C., Kay, T.S., Krentz, L. Nicotine alters striatal glutamate function and decreases the apomorphine-induced contralateral rotations in 6-OHDA-lesioned rats. *Exp. Neurol.* 175:257, 2002.

55. Mansvelder, H.D., Keath, J.R., McGehee, D.S. Synaptic mechanisms underlie nicotine-induced excitability of brain reward areas. *Neuron.* 33:905, 2002.

56. Erhardt, S., Schwieler, L., Engberg, G. Excitatory and inhibitory responses of dopamine neurons in the ventral tegmental area to nicotine. *Synapse.* 43:227, 2002.

57. Reid, M.S., Fox, L., Ho, L.B., Berger, S.P. Nicotine stimulation of extracellular glutamate levels in the nucleus accumbens: neuropharmacological characterization. *Synapse.* 35:129, 2000.

58. Fedele, E., Varnier, G., Ansaldo, M.A., Raiteri, M. Nicotine administration stimulates the *in vivo N*-methyl-D-aspartate receptor/nitric oxide/cyclic GMP pathway in rat hippocampus through glutamate release. *Br. J. Pharmacol.* 125:1042, 1998.

59. Domino, E.F., Minoshima, S., Guthrie, S.K. et al. Effects of nicotine on regional cerebral glucose metabolism in awake resting tobacco smokers. *Neuroscience.* 101:277, 2000.

60. Ghatan, P.H., Ingvar, M., Eriksson, L. et al. Cerebral effects of nicotine during cognition in smokers and non-smokers. *Psychopharmacology.* 136:179, 1998.

61. Shiah, I.S.Y.L. GABA function in mood disorders: an update and critical review. *Life.* 63:1289, 1998.
62. Epperson, C.N.O.M.S., Czarkowski, K.A., Gueorguieva, R., Jatlow, P., Sanacora, G., Rothman, D.L., Krystal, J.H., Mason, G.F. Sex, GABA, and nicotine: the impact of smoking on cortical GABA levels across the menstrual cycle as measured with proton magnetic resonance. *Biol. Psychiatry.* 57:44, 2005.
63. Magata, Y., Kitano, H., Shiozaki, T. et al. Effect of chronic (–) nicotine treatment on rat cerebral benzodiazepine receptors. *Nuclear Med. Biol.* 27:57, 2000.
64. Poindexter, E., Carpenter, R. The isolation of harmane and norharmane from tobacco and cigarette smoke. *Phytochemistry.* 1:215, 1962.
65. McIsaac, W., Estevez, V. Structure-action relationship of beta-carbolines as monoamine oxidase inhibitors. *Biochem. Pharmacol.* 15:1625, 1966.
66. Buckholtz, N., Boggan, W. Monoamine oxidase inhibition in brain and liver produced by b-carbolines: structure activity relationships and substrate specificity. *Biochem. Pharmacol.* 26:1991, 1977.
67. Rommelspacher, H., Nanz, C., Borbe, H., Fehske, K., Muller, W., Wollert, U. Benzodiazepine antagonism by harmane and other β-carbolines *in vitro* and *in vivo. Eur. J. Pharmacol.* 70:409, 1981.
68. Beuten, J.M.J., Payne, T.J., Dupont, R.T., Crews, K.M., Somes, G., Williams, N.J., Elston, R.C., Li, M.D. Single- and multilocus allelic variants within the $GABA_B$ receptor subunit 2 ($GABA_{B2}$) gene are significantly associated with nicotine dependence. *Am. J. Hum. Genet.* 76:859, 2005.
69. Karras, A., Kane, J. Naloxone reduces cigarette smoking. *Life Sci.* 27:1541, 1980.
70. Gorelick, D.A.R.J., Jarvik, M.E. Effect of naloxone on cigarette smoking. *J. Subst. Abuse.* 1:153, 1988.
71. Nemeth-Coslett, R., Griffiths, R.R. Naloxone does not affect cigarette smoking. *Psychopharmacology.* 89:261, 1986.
72. Sutherland, G., Stapleton, J., Russell, M., Feyerabend, C. Naltrexone, smoking behaviour and cigarette withdrawal. *Psychopharmacology.* 120:418, 1995.
73. King, A., Meyer, P. Naltrexone alteration of acute smoking response in nicotine-dependent subjects. *Pharmacol. Biochem. Behav.* 66:563, 2000.
74. Krishnan-Sarin, S., Meandzija, B., O'Malley, S. Naltrexone and nicotine patch in smoking cessation: a preliminary study. *Nicotine Tob. Res.* 5:851, 2003.
75. Hutchison, K., Monti, P., Rohsenow, D. et al. Effects of naltrexone with nicotine replacement on smoking cue reactivity: preliminary results. *Psychopharmacology.* 142:139, 1999.
76. Brauer, L., Behm, F., Westman, E., Patel, P., Rose, J. Naltrexone blockade of nicotine effects in cigarette smokers. *Psychopharmacology.* 143:339, 1999.
77. King, A. Role of naltrexone in initial smoking cessation: preliminary findings. *Alcohol. Clin. Exp. Res.* 26:1942, 2002.
78. Covey, L., Glassman, A., Stetner, F. Naltrexone effects on short-term and long-term smoking cessation. *J. Addict. Dis.* 18:31, 1999.
79. Wong, G., Wolter, T., Croghan, G., Croghan, I., Offord, K., Hurt, R. A randomized trial of naltrexone for smoking cessation. *Addiction.* 94:1227, 1999.
80. Almeida, L.E.P.E., Alkondon, M., Fawcett, E.P., Randall, W.R., Albuquerque, E.X. The opioid antagonist naltrexone inhibits activity and alters expression of alpha7 and alpha4beta2 receptors in hippocampal neurons: implications for smoking cessation programs. *Neuropharmacology.* 39:2740, 2000.
81. Boyadjieva, N.I.S.D. The secretory response of hypothalamic beta-endorphin neurons to acute and chronic nicotine treatments and following nicotine withdrawal. *Life Sci.* 61:PL59, 1997.
82. Berrendero, R., Kieffer, B., Maldonado, R. Attenuation of nicotine-induced antinociception, rewarding effects and dependence in mu-opioid receptor knock out mice. *J. Neurosci.* 22:10935, 2002.
83. Scott, D., Heitzeg, M., Ni, L., Domino, E., Zubieta, J. Endogenous opioid neurotransmission and tobacco smoking behavior: a PET study. Paper presented at the Society for Neuroscience, San Diego, CA, 2004.
84. Lerman, C.W.E., Patterson, F., Rukstalis, M., Audrain-McGovern, J., Restine, S., Shields, P.G., Kaufmann, V., Redden, D., Benowitz, N., Berrettine, W.H. The functional mu opioid receptor (OPRM1) Asn40Asp variant predicts short-term response to nicotine replacement therapy in a clinical trial. *Pharmacogenomics J.* 4:184, 2004.
85. Bever, K., Perry, P. Dexfenfluramine hydrochloride: an anorexigenic agent. *Am. J. Health Syst. Pharm.* 54:2059, 1976.
86. Fornal, C.R.M. Sleep suppressant action of fenfluramine in rats. I. Relation to postsynaptic serotonergic stimulation. *J. Pharmacol. Exp. Ther.* 225:667, 1953.

87. Berlin, I.S.S., Spreux-Varoquaux, O., Launay, J.M., Olivares, R., Millet, V., Lecrubier, Y., Puech, A.J. A reversible monoamine oxidase A inhibitor (moclobemide) facilitates smoking cessation and abstinence in heavy, dependent smokers. *Clin. Pharmacol. Ther.* 58:444, 1995.

88. Berlin, I., Said, S., Spreux-Varoquaux, O. et al. A reversible monoamine oxidase A inhibitor (moclobemide) facilitates smoking cessation and abstinence in heavy, dependent smokers. *Clin. Pharmacol. Ther.* 58:444, 1995.

89. Cornelius, J., Salloum, I., Ehler, J. et al. Double-blind fluoxetine in depressed alcoholic smokers. *Psychopharmacol. Bull.* 33:165, 1997.

90. Rausch, J., Fefferman, M., Ladisich-Rogers, D., Menard, M. Effect of nicotine on human blood platelet serotonin uptake and efflux. *Prog. Neuropsychopharmacol. Biol. Psychiatry.* 13:907, 1989.

91. Schievelbein, H., Werle, H. Mechanism of the release of amines by nicotine. *Ann. N.Y. Acad. Sci.* 142:72, 1967.

92. Berlin, I., Spreux-Varoquaux, O., Said, S., Launay, J. Effects of past history of major depression on smoking characteristics, monoamine oxidase-A and -B activities and withdrawal symptoms in dependent smokers. *Drug Alcohol Depend.* 45:31, 1997.

93. Racke, K., Schworer, H., Simson, G. Effects of cigarette smoking or ingestion of nicotine on platelet 5-hydroxytryptamine (5-HT) levels in smokers and non-smokers. *Clin. Invest.* 70:201, 1992.

94. Pfueller, S.L., Burns, P., Mak, K., Firkin, B.G. Effects of nicotine on platelet function. *Haemostasis.* 18:163, 1988.

95. Sparrow, D., O'Connor, G., Young, J., Rosner, B., Weiss, S. Relationship of urinary serotonin excretion to cigarette smoking and respiratory symptoms. *Chest.* 101:976, 1992.

96. Xu, Z., Seidler, F.J., Ali, S.F., Slikker, W., Jr., Slotkin, T.A. Fetal and adolescent nicotine administration: effects on CNS serotonergic systems. *Brain Res.* 914:166, 2001.

97. Batra, V., Patkar, A., Berrettini, W., Weinstein, S., Leone, F. The genetic determinants of smoking. *Chest.* 123:1730, 2003.

98. Patkar, A.A.G.R., Berrettini, W.H., Weinstein, S.P., Vergare, M.J., Leone, F.T. Differences in platelet serotonin transporter sites between African-American tobacco smokers and non-smokers. *Psychopharmacology.* 166:221, 2003.

99. Berlin, I., Said, S., Spreux-Varoquaux, O., Olivares, R., Launay, J.-M., Peuch, A. Monoamine oxidase A and B activities in heavy smokers. *Biol. Psych.* 38:756, 1995.

100. Kenny, P.J.F.S., Rattray, M. Nicotine regulates 5-HT$_{1A}$ receptor gene expression in the cerebral cortex and dorsal hippocampus. *Eur. J. Neurosci.* 13:1267, 2001.

101. Rasmussen, K.C.J. Nicotine withdrawal leads to increased sensitivity of serotonergic neurons to the 5-HT$_{1A}$ agonist 8-OH-DPAT. *Psychopharmacology.* 133:343, 1997.

102. do Prado-Lima, P.A.S.C.J., Ataufer, M., Oliveira, G., Silveira, E., Neto, C.A., Haggstram, F., Bodanese, L.C., da Cruz, I.B.M. Polymorphism of 5HT$_{2A}$ serotonin receptor gene is implicated in smoking addiction. *Am. J. Med. Genet. B.* 128B:90, 2004.

103. Shiffman, S.J.J., Kharallah, M., Elash, C.A., Gwaltney, C.J., Paty, J.A., Gnys, M., Evoniuk, G., DeVeaugh-Geiss, J. The effect of bupropion on nicotine cravings and withdrawal. *Psychopharmacology.* 148:33, 2000.

104. Lerman, C.R.D., Kaufmann, V., Audrain, J., Hawk, L., Liu, A., Niaura, R., Epstein, L. Mediating mechanisms for the impact of bupropion in smoking cessation treatment. *Drug Alcohol Depend.* 67:219, 2002.

105. Durcan, M.J.D.G., White, J., Johnston, J.A., Gonzales, D., Niaura, R., Rigotti, N., Sachs, D.P. The effect of bupropion sustained-release on cigarette craving after smoking cessation. *Clin. Ther.* 24:540, 2002.

106. David, S.P.N.R., Papandonatos, G.D., Shadel, W.G., Burkholder, G.J., Britt, D.M., Day, A., Stumpff, J., Hutchinson, K., Murphy, M., Johnstone, E., Griffiths, S.E., Walton, R.T. Does the DRD2-Taq1 A polymorphism influence treatment response to bupropion hydrochloride for reduction of the nicotine withdrawal syndrome? *Nicotine Tob. Res.* 5:935, 2003.

107. Hall, S., Reus, V., Munoz, R. et al. Nortriptyline and cognitive-behavioral therapy in the treatment of cigarette smoking. *Arch. Gen. Psychiatry.* 55:683, 1998.

108. Schneider, N.G.O.R., Steinberg, C., Sloan, K., Daims, R.M., Brown, H.V. Efficacy of buspirone in smoking cessation: a placebo-controlled trial. *Clin. Pharmacol. Ther.* 60:568, 1996.

109. Cohen, C.B.E., Galli, F., Lochead, A.W., Jegham, S., Biton, B., Leonardon, J., Avenet, P., Sgard, F., Besnard, F., Graham, D., Coste, A., Oblin, A., Curet, O., Voltz, C., Gardes, A., Caille, D., Perrault, G., George, P., Soubrie, P., Scatton, B. SSR591813, a novel selective and partial α4β2 nicotinic receptor agonist with potential as an aid to smoking cessation. *J. Pharmacol. Exp. Ther.* 306:407, 2003.

CHAPTER 4

Neurochemical and Neurobehavioral Consequences of Methamphetamine Abuse

Colin N. Haile, Ph.D.
Department of Psychiatry, Yale University School of Medicine, New Haven, Connecticut and VA Connecticut Healthcare System, West Haven, Connecticut

CONTENTS

Methamphetamine (METH) is a highly addictive and potent central nervous system (CNS) stimulant. Its rapid and escalating abuse in the U.S. has highlighted deficiencies in our understanding of the neurobiological mechanisms that underlie its powerful reinforcing effects. Availability of the drug facilitated through technological advances in synthesis and drug trafficking from other countries has also contributed to its rapid dissemination. According to the National Survey on Drug Use and Health, 12.3 million Americans have tried METH at least once, an increase of 40% from 2000 and 156% over 1996 numbers.[1] Although METH abuse was originally concentrated in the western part of the U.S. (Hawaii, California), recent statistics indicate a dramatic shift in its use to rural Midwest states. The National Clandestine Laboratory Database notes that the number of small-scale labs producing METH increased substantially in the Midwest (Illinois, Michigan, Ohio, Pennsylvania),[2] indicative of the redistribution of METH production and abuse in the U.S.[3] Availability of the drug, in turn, has resulted in substantial increases in substance abuse treatment admissions. Moreover, METH use is often associated with high-risk behaviors for transmitting HIV and other diseases. Because METH abuse has a profound impact on the health of the individual and society at large, it is paramount that we gain a better understanding of its effects on the human brain and its medical consequences.

4.1 MILITARY, MEDICAL USE, AND EVENTUAL ABUSE OF METH

METH is a derivative of amphetamine (AMPH) and both have many similarities in their effects on brain chemistry and behavior. They also share a common use history. AMPH was originally synthesized by Lazar Edeleanu in 1887 and again independently synthesized in 1927 by Gordon Alles.[4] It was eventually introduced commercially for the treatment of a myriad of ailments ranging from schizophrenia to hiccups.[5] AMPH has been used by the military to enhance concentration and vigilance ever since the Spanish Civil War. In World War II, American, German, British, and Japanese fighter pilots were administered the drug to stave off fatigue on long missions, a common use even today.[6] First synthesized by the Japanese pharmacologist Nagayoshi Nagai in the late 1800s, METH was also used during World War II to reduce soldier fatigue during military action and by civilians working in factories supporting the war effort. Similar to AMPH, METH was eventually sold over the counter in Japan beginning in 1941 as *Philopon* and *Sedrin*. Following the end of World War II, availability of METH increased further due to army surplus flooding the market. This initiated what has been called the "First Epidemic" (1945–1957) of METH abuse in Japan. Soon over half a million individuals were heavily abusing the drug, including 5% of the population between the ages of 16 and 25.[7] Strict laws were implemented in the 1950s to help deal with the problem. A "Second Epidemic" occurred in the 1970s when METH use increased among blue-collar workers, students, and housewives.[8] At present, METH abuse continues to be a serious problem in Japan and has remained the most popular illicit drug for the last 10 years.[9] In the U.S., underground METH labs appeared in California Bay Area in the 1960s. Recognizing the profitability of METH, motorcycle gangs began distributing the drug along the West Coast. METH abuse was so rampant it was the topic of a popular book at the time.[10] Drug enforcement crackdowns on gang activity and limitation of precursor chemicals for the synthesis of METH quelled the distribution to a certain degree. At present, however, the bulk of the West Coast drug supply — and the Midwest — appears to be coming from Mexican "super labs" with a minor percentage produced by small-scale establishments or so-called "mom-and-pop" laboratories.[3]

4.2 CHARACTERISTICS AND PATTERNS OF USE AND ABUSE

METH has been available legally in the U.S. for many years as a therapeutic drug (trade name Desoxyn) used to treat obesity and attention-deficit/hyperactivity disorder (ADHD). Illegal street forms are commonly known as "speed" or "meth," which can be self-administered via injection, smoking, nasal inhalation ("snorting"), or oral ingestion. In its highly pure smokable form it is referred to as "crystal," "ice," "crank," or "glass." When ingested, the lipophilic compound efficiently penetrates the CNS,[11] increasing concentrations of monoamines (particularly dopamine, DA) through multiple mechanisms (see below).[12] The intensity of the "high" and mood alteration produced from METH ingestion is route dependent. Smoking and injection result in an almost immediate euphoria or "rush," whereas the effect is less intense and rapid when administered via "snorting" (effects felt within 3 to 5 min) or oral ingestion (15 to 20 min).[1] The half-life of METH is an impressive 10 h in humans (compared to 90 min for cocaine).[13-15] METH abusers tend to self-administer the drug in a "binge and crash" pattern. "Binges" or "runs" may last for 1 to 3 days or more followed by a period of abstinence. Because of the long half-life of METH in humans, "binge" administration results in successive accumulation of residual drug in the system. Tolerance to many — but not all — of the peripheral and central effects of METH occurs almost immediately.[16,17] Acute METH intoxication results in powerful stimulation of the sympathetic nervous system resulting in mydriasis (pupil dilation), hypertension, tachycardia, diaphoresis, and hyperthermia. The reinforcing or positive effects of acute administration include euphoria, increased energy, heightened attentiveness, hypersexuality, and decreased anxiety.[18,19] Upon withdrawal from METH the individual is said to "crash," which is discernible by the presence of depression, anhedonia, irritability, anxiety, fatigue, hypersomnia, poor concentration, intense craving, and aggression.[19-22] In certain respects, symptoms are more intense and distinguishable from amphetamine and cocaine withdrawal.[21,23,24] Individuals who have been consuming METH frequently and for long periods of time show severe psychiatric disturbances or METH "psychosis," which has many characteristics in common with schizophrenia.[9,25-28]

4.3 NEUROCHEMISTRY OF METH: MECHANISMS OF ACTION AND REINFORCEMENT

4.3.1 Dopamine

METH is similar to other drugs of abuse in that its reinforcing effects are mediated through multiple sites and mechanisms in the brain. It is well established that drugs of abuse — and natural reinforcers such as food — exert their effects, in part, by activation of the mesolimbic DA system.[29-31] This system consists of DA cell bodies in the ventral tegmental area (VTA) and their forebrain terminals in the prefrontal cortex (PFC) and nucleus accumbens (NAC). Drugs abused by humans evoke DA release in the PFC and NAC, the latter a crucial brain substrate that mediates the reinforcing or addictive aspects of drugs of abuse.[32] It is hypothesized that DA release in these areas increases the saliency or attractiveness of rewarding stimuli contributing to the addiction process.[33] Addictive drugs including METH are self-administered under controlled conditions[34,35] and activate mesolimbic DA in humans.[36] Lesions at different loci of this system alter the behavioral effects of drugs of abuse in animals.[37,38] Released DA from terminal regions subsequently binds to a number of DA receptor subtypes such as D_1-like (D_1, D_5) or D_2-like (D_2, D_3, D_4), which are classified based on molecular and pharmacological characteristics. DA neurotransmission is then terminated by sequestration of the transmitter into the presynaptic neuron through the dopamine transporter (DAT).[39,40] Depending on the behavioral paradigm, drugs that block DA receptors alter the behavioral effects of drugs of abuse to varying degrees.[30]

In animals, repeated intermittent psychostimulant administration (i.e., cocaine, AMPH, METH) enhances locomotor behavior over time. This phenomenon is referred to as behavioral sensitization. Sensitization is considered a key characteristic in the development of drug addiction and believed by some to be a model of psychosis or schizophrenia.[41-43] Induction of sensitization appears to be related to enhancement of DA neurotransmission in the mesolimbic DA system, although other neurotransmitters such as glutamate (GLU), serotonin (5-HT), and norepinephrine (NE) are involved.[44] Drugs that block behavioral sensitization may have pharmacotherapeutic potential. Another important concept in addiction is tolerance. Tolerance is the decrease in behavioral response to the same dose of the drug over time and most likely plays a role in the increasing amounts of ingested drug over time by drug addicts.[45]

Similar to cocaine and AMPH, METH has strong effects at the DAT, which are likely responsible for its potent addictive properties. In fact, cocaine's reinforcing effects are related to its ability to enhance extraneuronal DA concentrations by blocking the DAT.[46] Likewise, AMPH increases DA levels by primarily reversing the DAT and inducing transmitter release into the extracellular space.[47,48] Strong evidence supporting this assumption comes from finding that neither cocaine nor AMPH is effective in genetically modified mice lacking this transporter.[49] Acute administration of METH also potently increases DA concentrations in reward circuitry[50–52] via an exchange diffusion mechanism independent of neuronal depolarization[53] and by redistributing cytosolic DA to areas in the neuron for quick fusion and discharge.[54,55] Systemic treatment for 7 days with METH enhances the response of mesolimbic VTA cell body neurons to subsequent administrations of the drug. This effect is antagonized with Ca^{2+} channel blockers.[56] A similar treatment regimen (5 days) results in hypersensitivity of VTA neurons altering the maximum amplitude and the ED_{50} value of D_2 receptor-mediated hyperpolarization.[57] Hyperpolarization (i.e., inhibition) of DA neurons in the VTA may be a compensatory mechanism engaged to decrease excessive DA release in the NAC. However, an attenuation of DA release in this terminal region results in sensitized DA receptors, which is perhaps also compensatory. Consistent with this notion, *in vitro* intracellular recording in brain slices from rats pretreated with METH shows supersensitized D_1 receptor-mediated hyperpolarizations in the NAC.[58]

The exact mechanisms responsible for the ability of METH to increase extracellular DA are fairly well delineated and are, in part, due to facilitation of DA discharge and inactivation of the DAT. For instance, the release of massive quantities of DA facilitates the formation of reactive oxygen species via auto-oxidation of DA[59] that, in turn, inactivates DAT.[60,61] Inactivation of DAT increases synaptic DA by preventing reuptake into the presynaptic neuron. Other reactive species such as superoxide or peroxynitrite also inactivate DA by oxidization, transforming it into highly reactive DA quinones that can also compromise DAT function.[62,63] Indeed, experiments show that acute and chronic administrations of METH cause a rapid and reversible decrease in the DAT.[64,65] Remarkably, a single METH injection dose-dependently decreases [^{3}H]dopamine uptake into striatal synaptosomes 1 h after treatment, suggesting rapid deleterious effects on DAT function.[64] DAT inactivation is blocked by depleting DA using the tyrosine hydroxylase inhibitor α-methyl-*p*-tyrosine[66] and by pretreatment with D_1 and D_2 antagonists and DAT blockers.[66–68] This suggests that abnormally high levels of DA evoked by METH may be the causative agent underlying DAT inactivation. Yet, studies also demonstrate that the potent hyperthermic effects of METH aid in enhanced production of reactive oxygen species that may further contribute to the inactivation of DAT.[63,69] Neutralizing METH-induced increases in body temperature[66] blocks its effects on the DAT, suggesting that inactivation involves a multicomponent process.

As DA is taken back up into the presynaptic neuron, it is sequestered into synaptic vesicles and repackaged for storage and subsequent re-release, a process mediated by the vesicular monoamine transporter (VMAT-2) in monoaminergic neurons. Once inside the neuron, DA is protected against oxidation, which could produce reactive oxygen species implicated in DAT inactivation.[70] Indeed, acute and multiple administrations of METH rapidly and persistently (up to 24 h) alter (within 60 min) vesicular [^{3}H]DA uptake as assessed in vesicles purified from striatum.[71,72] Pretreatment with the DA D_2 antagonist eticlopride but not the D_1 antagonist (SCH23390) prevents decreases in vesicular DA uptake by METH.[73] These data implicate D_2 receptors in METH-induced decreases of VMAT-2. As mentioned previously and consistent with other transmitter-releasing compounds,[47] METH administration redistributes vesicles within the nerve terminal for immediate release, interestingly, in a fashion opposite to that of cocaine.[74] The distinctive difference between the two drugs may contribute to differences in their neurotoxic and behavioral profiles. Subcellular fraction preparations from striatum in rats analyzed at 24 h after METH administration show reduced overall VMAT-2 protein suggesting actual degradation occurs.[75] Taken together, VMAT-2 inactivation would hypothetically lead to increased cytosolic DA levels and potential formation of reactive oxygen species by auto-oxidation or by monoamine oxidase (MAO) leading to neurotoxicity,[76] a prominent feature of chronic METH consumption.[77]

4.3.2   Serotonin

In a number of ways, the effects of METH on serotonin (5-HT) are similar to those on DA. For example, repeated METH injections increase hippocampal (250%, over controls)[78] and nucleus accumbal (900%) extracellular 5-HT levels.[52] Long-lasting deficits in 5-HT metabolite parameters occur in the striatum, hippocampus, and frontal cortex in response to multiple administrations of METH.[79,80] A single high dose of METH (15 mg/kg) decreases tryptophan hydroxylase — the rate-limiting enzyme in 5-HT synthesis — in the NAC and caudate.[81] Previous studies confirm these effects and posit that — similar to DA — inactivation may be caused by reactive oxygen species formed inside 5-HT terminals oxidizing the enzyme and causing deleterious effects to the neuron.[79,82,83] Acute and multiple injections of METH (10 mg/kg) result in reversible decreases in 5-HT transporter (SERT) function *in vivo*,[68,84] whereas high doses of fenfluramine, cocaine, or methylphenidate do not.[85] High (15 mg/kg) but not low (7.5 mg/kg) doses of METH administered repeatedly reduce the binding of [³H]cyanoimipramine ([³H]CN-IMI) to serotonin uptake sites assessed by quantitative autoradiography.[86] Similar to DA, studies have shown that inactivation of SERT may also be due to the production of reactive oxygen species such as the endotoxin tryptamine-4,5-dione, a by-product of oxidized 5-HT.[87] Acute METH administration blocks SERT function in the striatum but not in the hippocampus. This effect appears to be mediated through DAergic pathways and partly by the hyperthermic effects of METH. Decreasing METH-induced increases in body temperature, depleting striatal DA with α-methyl-p-tyrosine, or pretreatment with D_1 and D_2 antagonists (SCH23390 and eticlopride) blocks the ability of METH to decrease SERT activity in the striatum but not in the hippocampus.[88] These results suggest that the action of METH on SERT localized in the striatum is predominantly mediated through DA and that hyperthermia also plays a role. Hippocampal changes appear to be dependent on 5-HTergic pathways. In addition, like DAT blockers, SERT blockers (citalopram and chlorimipramine) are also neuroprotective.[82,89]

4.3.3   Norepinephrine

Evidence from human and animal studies highlights a unique role for norepinephrine (NE) in the neurobiological effects of METH. For example, METH increases extracellular NE divergently in the caudate and hippocampus of rodents as measured by microdialysis.[90] Depletion of NE with the selective neurotoxin N-(-2-chloroethyl)-N-ethyl-2-bromobenzylamine (DSP-4) (50 mg/kg) significantly enhances METH-induced striatal DA depletion in rodents.[91] Pharmacological blockade of NE with clonidine, a drug that shuts down NE release via presynaptic α_2 adrenergic autoreceptors, potentiates METH-induced effects, whereas blockade of α_2 with antagonists (e.g., yohimbine), which enhances release, reduces the drug's deleterious effects.[91] These results suggest that NE may help attenuate alterations in neurochemistry attributed to DA. An early study also demonstrated that METH-induced increases in tryptophan hydroxylase activity are blocked with the NE antagonist propranolol indicating NE and 5-HT coordinate in some unknown way.[92] Unlike DAT and SERT, however, NET appears to be less vulnerable to the adverse effects of METH. METH treatment does decrease NET activity in synaptosomes; however, these changes are due to a direct effect of the drug on the transporter and not by indirect inactivation via reactive oxygen species seen with DA and 5-HT. Indeed, the aberrant effects on NETs can be reversed by simply rinsing the *in vitro* preparation of residual METH.[93] In addition, high doses of METH administered over a 2-week period result in depletion of DA and 5-HT but not NE in nonhuman primate brain.[94] Similarly, single or repeated METH administration reduces many neurochemical metabolic parameters associated with DA and 5-HT but not NE in the striatum-accumbens and thalamus-hypothalamus in mice.[95] Although it appears that NE plays a minimal role in the action of METH on the brain, there is evidence NE may be important. An *in vitro* study has recently shown that oral doses of psycho-stimulants, including METH, which produce subjective effects in humans, correlate with their potency to release NE, not DA or 5-HT,[96] and prazosin, an α_1 adrenergic antagonist, blocks cocaine-induced reinstatement in an animal model of relapse.[97] Moreover, human METH abusers who develop spontaneous recurrence of METH psychosis show markedly elevated NE plasma levels, indicating that this neurotransmitter may be of prime importance.[98]

4.4 NEUROBEHAVIORAL AND NEUROPHARMACOLOGICAL EFFECTS OF METH: PHARMACOTHERAPEUTIC TARGETS

Experiments in rodents and other animals allow us to closely examine the complex interplay between the drive or motivation to consume addictive drugs and behavior. This information has helped determine the neurobiological substrates that are responsible for the reinforcing or addictive effects of drugs of abuse.[99] Accumulating evidence suggests that the robust abnormal drug-seeking behavior seen in the addicted state is due, in part, to drug-induced alterations in neural sensitivity, neurotransmitter levels, and neural plasticity that is heavily embedded in learning.[100,101] Teasing out the mechanisms behind drug-induced alterations in brain proteins in areas that mediate these addictive states is of prime importance. Likewise, drugs that reverse or block these changes may serve as useful pharmacotherapies. Therefore, the effects of METH in the context of motivation and drug-induced changes in reinforcement-related brain circuits and possible drug therapy targets are reviewed below.

4.4.1   Dopamine's Vital Contributions to METH-Induced Reinforcement

Similar to cocaine and AMPH, acute administration of METH (2 mg/kg) potently increases DA in the NAC 1000% over baseline levels.[102,103] METH administration also leads to the develop-ment of behavioral sensitization[95,104] that is heavily dependent on dose and drug regimen.[105,106] As a testament to the reinforcing properties of METH, and like other psychostimulants, the drug is readily self-administered across a number of species.[107–112] In fact, self-administration of METH in combination with other drugs of abuse such as heroin makes it even more reinforcing.[113] Consistent with the action of METH on the DA system, drugs that modulate DA in one way or another alter METH-induced behaviors. For example, co-treatment with either a D_1 (SCH23390) or a D_2 (YM-09151-2) antagonist blocks the development and expression of METH-induced (4 mg/kg) behavioral sensitization in rats over 14 days of treatment.[104] Correspondingly, Witkin et al.[104a] demonstrated that pretreatment with the highly selective D_1 antagonist SCH39166 or the D_2 antagonist spiperone blocks the behavioral activation of METH (0.3 mg/kg) in mice. METH-induced behavioral sensitization in animals is used as a model of psychosis and drugs that antagonize this effect may be useful in treating disorders such as schizophrenia.[43] Particular attention has been focused on the D_4 receptor subtype when it was discovered that the atypical antipsychotic clozapine blocks this receptor among its many other actions.[114] Pretreatment with a selective D_4 antagonist (NRA 0160) blocks METH-induced hyperactivity in mice to a similar degree to that of clozapine.[115] It is unknown if clozapine would prove a useful treatment for METH abuse.

Given that METH, and other psychostimulants, readily bind and modulate SERT, DAT, and NET to varying degrees, compounds acting on these transporters in unison may prove therapeutic. Indatraline, a compound that binds to 5-HT, NET, and DAT, was recently shown to inhibit METH-induced DA release *in vitro*.[116] In a rat model of relapse, priming injections of indatraline marginally reinstated previously extinguished cocaine-seeking behavior (lever pressing for drug) as measured by self-administration, yet failed to alter overall drug intake.[117] Along these lines, the 3-phenyltro-pane analogue RTI 111 that is marginally selective for the DAT, yet also has proclivity for the other transporters, increases the potency of self-administered METH in nonhuman primates.[118] Although counterintuitive, other drugs that enhance the effects of psychostimulants such as cocaine to the point at which they are aversive have proven efficacious.[119] However, RTI 111 is readily self-administered in a manner similar to cocaine and thus may possibly be abused itself. Other drugs that are more selective for the DAT, such as GBR12909, have been tested. GBR12909 inhibits AMPH transport into striatal synaptosomes suggesting that it could attenuate the behavioral effects of its cousin METH.[120] Similar to RTI 111, however, GBR 12909 co-treatment with METH potentiates the discriminative stimulus effects of METH in rats, a behavioral model that tests the subjective effects of drugs.[121] Moreover, priming injections of GBR 12909 reinstated previously extinguished cocaine-seeking behavior as measured by self-administration.[117] The results mentioned above emphasize the fact that experimental results attained *in vitro* are poor predictors for how the drug will behave *in vivo*. Nevertheless, a recent study with a long-lasting version of GBR12909 shows promise as a treatment for METH addiction in preclinical models.[52] Whether compounds targeting the DAergic system will produce optimal treatments for METH remains to be seen.

4.4.2 Serotonin Modulation of the Reinforcing Effects of METH

Aside from the known contribution of DA, preclinical studies indicate 5-HT plays a role in the reinforcing effects of drugs of abuse. For instance, mice lacking 5-HT_{1B} receptors are hypersensitive to the behavioral activating effects of cocaine.[122] Lesions of forebrain 5-HTergic tracts increase amphetamine self-administration suggesting that 5-HT regulates DA-mediated effects to a degree.[123] Indeed, it is well known that stimulating 5-HT by various means can augment DA neurotransmission.[124,125] Although the contribution of 5-HT in the effects of METH is not fully known, METH does indeed potently activate the 5-HTergic system,[84] enhancing release[126] and increasing extracellular levels in brain.[90] Munzar et al.[127] demonstrated that the powerful 5-HT-releaser fenfluramine initially decreases METH self-administration. However, due to unknown mechanisms, tolerance developed to this effect after repeated dosing.[127] Results from drug discrimination experiments in that same study found that various 5-HT compounds targeting a number of receptor subtypes modulate and/or generalize to the discriminative stimulus effects elicited by METH.[127] These results are consistent with other studies demonstrating the modulatory effects of 5HT on METH-induced behaviors. For example, pretreatment with 5-HT_{1A} (NAN-190), $5\text{-HT}_{1B/1D}$(methiothepin), and 5-HT_{2C} (mianserin) antagonists attenuates the acute locomotor stimulating effects of METH, whereas $5\text{-HT}_{2A/2B}$ (methyserigide) and 5-HT_3 (ondansetron) antagonists potentiate the METH effects.[128,129] The mechanisms that underlie the ability of different 5-HTergic compounds to divergently alter the behavioral effects of METH are unknown. Taken together, however, these data suggest that 5-HT likely plays more of a modulatory role than that of DA in METH-induced behaviors.[130] Drugs acting on this system may prove useful treatments especially for abnormalities in mood and aggression associated with METH withdrawal.

4.4.3 Glutamate and METH-Induced Behaviors

Glutamate (GLU) is the most abundant neurotransmitter in the brain and clearly has an important position in addiction. Indeed, GLU is essential in psychostimulant-induced sensitization[131] and reinforcement[132,133] by possibly altering DA neurotransmission in the PFC.[134] Remarkably, mice genetically lacking the metabotropic GLU receptor GluR5 are immune to the locomotor and reinforcing effects of cocaine.[135] Compounds that block this receptor also attenuate the reinforcing effects of other drugs of abuse.[136] Although METH and AMPH are similar and share common biochemical and behavioral effects, METH administration increases GLU levels in the PFC to a greater extent compared to AMPH.[103] The direct consequences of this difference in GLU-releasing ability are not known but may be important in terms of drug-associated neuroplasticity and treatment.

Consistent with the notion that GLU is important in the behavioral effects of METH, compounds that block AMPA-type glutamatergic receptors (NBQX) [137,138] or NMDA receptors (NPC 12626) decrease METH-induced locomotion. However, only high doses of NPC 12626 that disrupt normal locomotor behavior are effective,[138] indicating that the METH effects are most likely mediated largely through the AMPA receptor subtype. Similarly, drugs that facilitate removal of METH-induced increases of GLU from the extracellular space block its rewarding effects[139] as measured by a place conditioning paradigm.[140] The clinical implications for METH-induced increases of GLU in the context of drug abuse are not known. However, current evidence suggests individuals with obsessive-compulsive behavior or disorder (OCD) show hyperglutamatergic activity in the PFC.[141] Obsessive-compulsive behavior is akin to uncontrollable drug-seeking and individuals with OCD have an increased likelihood of drug abuse.[142]

4.4.4   Novel Therapeutic Targets for METH Addiction

A number of studies have tested compounds that home in on other novel neurotransmitter systems and reveal important clues to the action of METH. Initially classified as an opioid receptor, sigma (σ) receptors (sigma-1 and sigma-2) have been implicated in a variety of psychiatric disorders including depression, anxiety, schizophrenia,[143,144] and, more recently, psychostimulant addiction.[145] Interestingly, sigma receptors are strategically localized in the nucleus accumbens and other areas within limbic circuitry.[145] Studies have demonstrated that psychostimulants bind to sigma receptors[146] and sigma (1) antagonists block many of the behavioral effects of cocaine and AMPH.[147,148]

Like other psychostimulants, *in vitro* binding studies show that METH also preferentially binds to sigma-1 receptors and pretreatment with sigma-1 receptor antagonists, such as BD1063 or BD1047, attenuates its acute behavioral activating effects.[148] Similarly, antisense oligodeoxynucleotides aimed at sigma-1 receptors, acting as a molecular antagonist, attenuate the locomotor-stimulating effects of METH. Evidence shows that psychostimulants either increase the number or sensitivity of sigma receptors *in vivo* and this also appears to be the case for METH. Indeed, rats previously sensitized to METH are significantly more responsive to the sigma receptor agonist (+)3-(3-hydroxyphenyl)-N-(1-propyl)piperidine ((+)-3-PPP).[149] Repeated administration of METH increases binding of the sigma ligand [^{3}H](+)pentazocine in a number of brain areas in rodents.[150] Sigma-1 receptors are also upregulated (protein and mRNA) in rats that self-administer but not in those that passively received METH.[151] Most importantly, sigma-1 antagonists block METH-induced behavioral sensitization.[152,153] The exact mechanism through which sigma-1 receptors are responsible for neutralizing the action of METH is unknown. However, experiments show that sigma-1 receptors mediate cellular restructuring via cholesterol and cytoskeletal trafficking from the endoplasmic reticulum to the plasma membrane and nucleus.[154,155] It is likely, then, that sigma-1 receptors may be involved in psychostimulant-induced neuroplasticity related to uncontrollable drug intake and by blocking these receptors may interrupt this process.[156] Although details are still emerging, these studies suggest a crucial role for the sigma receptor in the behavioral effects of METH and may prove a useful drug treatment target.

Early studies provided support for an alkaloid (*ibogaine*) found in the root bark of the African shrub *Tabernanthe iboga* having anti-addictive properties. Concerns of toxicity associated with *ibogaine* led to the development of the *iboga* alkaloid congener 18-MC (18-methoxycoronaridine).[157] Experiments in rodents show that 18-MC *enhances* METH-induced locomotion[158] and reduces METH self-administration.[159] These results are consistent with recent reports showing that disulfiram, a clinically efficacious compound for the treatment of cocaine addiction,[119] enhances the development and expression of cocaine-induced behavioral sensitization in rats.[160] Binding studies *in vitro* determined that *ibogaine* and 18-HC act as potent antagonists at $\alpha_3\beta_4$ nicotinic acetylcholine receptors with less potency seen at $\alpha_4\beta_2$, NMDA, or 5-HT$_3$ receptors.[161] Drugs such as mecamylamine and dextromethorphan that also antagonize $\alpha_3\beta_4$ block METH self-administration, lending further support for this receptor as a novel therapeutic target.[161,162] Indeed, lobeline, the lipophilic alkaloid obtained from the herb *Lobelia inflata*, also blocks $\alpha_3\beta_2$ and $\alpha_4\beta_2$ nicotinic neuronal receptors and has demonstrated great potential as a possible treatment for psychostimulant abuse. Lobeline pretreatment inhibits METH-induced locomotion, blocks the discriminative stimulus cue elicited by METH,[163] and decreases self-administration in rats.[164] Surprisingly, increasing the dosage of METH does not surmount the antagonism by lobeline suggesting good pharmaco-therapeutic potential. How lobeline is able to block the powerful reinforcing effects of METH is unknown, although studies indicate that the ability of lobeline to block the METH effects is not due to preventing METH-induced elevations of DA, but more likely due to its ability to prevent decreases in VMAT-2 and induction of hyperthermia.[75] Yet, lobeline also interacts with DAT[165] and increases 5-HT release that may involve SERT and contribute toward its anti-addictive effects.[166] Analogues of lobeline for the treatment of psychostimulant abuse are being developed.[167] Other preclinical studies testing possible novel treatments for METH abuse have targeted GABA,[168] cannabinoid,[169,170] and histamine receptors.[171] As has been attempted for cocaine addiction, a monoclonal antibody vaccine against METH is also in the developmental phase.[172]

4.4.5 METH-Induced Alterations in Intracellular Messenger Systems Related to Reinforcement

Recent research has focused on alterations in intracellular messenger systems and regulation of gene expression in response to drugs of abuse.[173–175] Similar to changes at the neurotransmitter level, molecular alterations occur in areas of the brain that mediate the reinforcing aspects of drug addiction and are long-lasting.[176] Drug-induced alterations are well thought of as a form of neural plasticity.[156] This neural plasticity occurs in response to modified gene expression that eventually leads to changes in neurotransmitter–receptor dynamics. In fact, every major drug of abuse produces long-term neuroplasticity in, for example, the VTA.[177] Understanding these alterations at the cellular level will inevitably improve our understanding of the underlying neural adaptations that govern addiction.

The first intracellular pathway to be thoroughly examined in the context of drug abuse was the cAMP/PKA/CREB cascade.[178] Neurotransmitters or drugs that activate D_1 receptors facilitate (acting through $G_{\alpha s}$ stimulatory G proteins), whereas neurotransmitters or drugs that activate D_2 (acting through $G_{\alpha i}$ inhibitory G proteins) decrease the formation of cyclic adenosine 3,5-monophosphate (cAMP). cAMP, in turn, affects cAMP-dependent protein kinase (PKA). The formation of cAMP is dependent on adenylyl cyclase and is degraded by various phosphodiesterase enzymes in the cytoplasm.[179] Drugs of abuse alter the dynamics of this intracellular messenger system. For example, repeated psychostimulant administration results in decreases in inhibitory G proteins ($G_{\alpha i}$) linked to D_2 receptors,[180,181] and elevated tyrosine hydroxylase[181–183] in the VTA. A number of persistent neuroadaptations are seen in the NAC in response to drug exposure. These include psychostimulant-induced supersensitivity of D_1-mediated effects,[184] decreased levels of $G_{\alpha l}$, but no effect on Gs G proteins,[185,186] increased adenylyl cyclase, cAMP-dependent protein kinase (PKA),[185] and immediate-early gene expression of fos-associated proteins such as ΔFosB.[187–189] Enhancing cAMP activity in the VTA potentiates psychostimulant sensitization and inactivation of PKA blocks this effect.[190] Infusion of cAMP analogues, Rp- and Sp-cAMPS, bilaterally into the NAC, that block and facilitate PKA, respectively, induce and prevent relapse of cocaine-seeking behavior.[191] Of primary importance is recent work on cAMP-response element-binding protein (CREB), a transcription factor localized in the nucleus that plays a crucial role in gene expression and plasticity-associated events.[192,193] CREB has been implicated in a number of behavioral processes, in particular, drug-induced sensitization[194] and reinforcement.[175,195,196]

Elevating cAMP/PKA levels in the NAC enhances, whereas blocking PKA attenuates, the expression of cocaine-induced locomotor sensitization.[197] Likewise, recent reports demonstrate that the behavioral activating effects of METH can be antagonized by indirectly increasing cAMP levels with rolipram, a selective inhibitor of cAMP-specific phosphodiesterase 4 that degrades cAMP.[198] Co-treatment with systemic rolipram (4 mg/kg) blocks METH-induced activation in rats following a sensitizing treatment regimen (4 mg/kg × 5 days, 1 week withdrawal, then a 2 mg/kg METH challenge). Rolipram does not alter METH-induced increases in extracellular levels of DA in the striatum suggesting that the antagonism of the behavioral effects of METH were most likely due to increases in cAMP.[199] These data are in complete agreement with Mori et al.,[199a] showing rolipram co-treatment blocks METH and morphine's locomotor activating effects but not those elicited by phencyclidine. The authors found that very high doses of rolipram (10 mg/kg) only partially attenuated SKF81297-induced (D_1 agonist) locomotion. Therefore, these data suggest that METH effects were likely blocked by increasing cAMP through inhibitory D_2 receptors. Post-mortem findings in METH abusers support this notion (see below).

Activated through the D_1 receptor pathway, DA and cAMP-regulated phosphoprotein 32 kDa (DARPP-32) is a substrate for PKA found in the striatum and is involved in molecular adaptations that occur in response to drugs of abuse.[200] Phosphorylation by PKA converts DARPP-32 into an efficient inhibitor of PP1 (protein phosphatase-1). Consistent with the known fact that psychostimulants alter cellular responses acutely and long-term, PP1 has been shown to modulate AMPA channels involved in neuronal plasticity.[201–203] Once activated by PKA, however, DARPP-32 then affects a variety of downstream physiological effectors.[200] Studies in genetically modified mice lacking DARPP-32 show altered responses to psychostimulants.[204,205] Like cocaine, acute METH administration (20 mg/kg) increases DARPP-32 immunoreactivity and phosphorylation of various residues associated with GLU receptor subtypes in the neostriatum in wild-type but not in DARPP-32 knockout mice.[206] This effect was also shown *in vitro* and *in vivo* in the striatum of rats sensitized to METH.[207,208] DARPP-32 is also phosphorylated by a cyclin-dependent kinase (cdk5) that reverts the protein into a PKA inhibitor.[209] Interestingly, intra-NAC injections of roscovitine, a cdk5 inhibitor, attenuates METH-induced locomotor sensitization.[208] Moreover, recent evidence has connected cdk5 with ΔFosB, a transcription factor implicated in long-term adaptations to drugs of abuse.[210] However, whether METH induces the expression of ΔFosB is not known. METH also affects ARPP-21, a cAMP-regulated phosphoprotein of 21 kDa that is also phosphorylated by PKA and enriched in limbic structures.[211] Acute administration of METH or cocaine increases ARPP-21 phosphorylation in rodents.[212] What role these proteins play in METH-induced behavioral effects such as reinforcement is unclear.

Intracellular signaling is heavily dependent on Ca^{2+}, and drug-induced alterations could have profound effects on normal neuronal function. Indeed, chronic METH decreases kinases associated with Ca^{2+} such as Ca^{2+}/calmodulin (CaM)-dependent protein kinase II (CaM-kinase II) specifically in the VTA-NAC pathway that is blocked by the D1 antagonist SCH23390 and MK801, a GLU antagonist.[213,214] Other Ca^{2+}-associated proteins are also affected by METH, for example, calmodulin, a calcium-binding protein also implicated in the effects of other drugs of abuse.[215] Similar to the effects seen on CaM-kinase II, chronic METH (4 mg/kg × 14 days, 28-day withdrawal, and a 4 mg/kg challenge) significantly decreases calmodulin mRNA in the NAC and VTA. Comparable decreases have been seen in calcineurin in the striatum of rats sensitized to METH.[207] It is not known, however, whether these decreases affect neuronal function in a manner associated with drug sensitization or reinforcement. However, reduced activity of Ca^{2+} proteins involved in intracellular trafficking would undoubtedly have effects on several substrate proteins that are important for proper neuronal functioning.

Experiments conducted *in vitro* using immunofluorescence and mobility shift assays reveal that acute application leads to accumulation of METH in cytosol and vesicular compartments (4 to 6 h) and eventual translocation into the nucleus. In the nucleus, METH increases activator protein-1 (AP-1) and CREB DNA binding activity.[216,217] Pre-incubation with an anti-METH antibody prevents the enhancement of these DNA-binding proteins.[218] Experimental evidence shows that METH-evoked enhancement of AP-1 and CREB binding but not of other transcription factors (NF-KB, SP-1, STAT1, STAT3) is dose-dependent and is apparent in brain areas involved in reinforcement such as the frontal cortex and hippocampus.[219] METH (4 mg/kg) administration for 2 weeks with a 1-week interval and a final challenge — a treatment that produces drug sensitization — results in significant increases in cFos, CREB, and pCREB (phosphorylated form of CREB) immunoreactivity in rat striatum.[220] Animals learn preferences for places (place preference) where they have previously experienced a reward. Drugs that are more rewarding induce robust place preferences. In contrast, drugs that are not rewarding may produce aversion.[221] Recent studies show that CREB plays a primary role in the rewarding effects of psychostimulants. For example, Carlezon et al.[221a] demonstrated using viral transfer techniques that overexpression of CREB in the NAC makes cocaine aversive, whereas blocking CREB enhances the rewarding attributes of the drug. Whether modulating CREB in the NAC will alter METH-induced reward is not known.

While the role of molecular adaptations in response to drugs of abuse affecting the cAMP/PKA/CREB signaling cascade has been thoroughly investigated, less attention has been paid to other intracellular pathways. For example, the mitogen-activated protein kinase (MAPK) pathway plays an important role in cell growth, differentiation, proliferation,[222] and neural plasticity associated with learning and memory.[223] Evidence suggests that drug-induced maladaptive forms of neural plasticity in areas of the brain involved in reward learning[101,156] may underlie the uncontrolled drug-seeking and drug intake seen in addiction.[224] For example, changes in plasticity-related genes in response to METH include tissue plasminogen activator,[225] activity-regulated cytoskeleton-associated protein,[226] synaptophysin, and stathmin[227] and MAP kinase phosphatases.[228] A number of other gene-products associated with this pathway are altered in METH-induced sensitized animals.[229] METH-evoked expression of genes involved in neuronal remodeling in limbic brain areas could contribute to drug-reward processes. For example, a number of studies in rodents implicate MAPK pathway in psychostimulant-induced sensitization and reward learning.[230–232] Consistent with these results, Mizoguchi et al.[232a] provide definitive evidence involving MAPK and METH-induced reward conditioning. Results show hyperphosphorylation of MAPK/ERK1/2 in the NAC and striatum, but not in other areas in rats that had previously undergone METH-induced place conditioning. Pretreatment with both D_1 (SCH23390) and D_2 (raclopride) antagonists and PD98059 (a selective MAPK inhibitor) directly infused into the NAC blocks METH-induced place preference conditioning and ERK1/2 activation. This suggests a critical involvement of the MAPK/ERK signaling cascade in METH-evoked reward learning.

4.4.6   METH-Induced Alterations in Intracellular Messenger Systems in Humans

Results from post-mortem human studies addressing METH-induced changes in receptors and intracellular messenger systems are generally in line with changes seen in animal studies (Figure 4.1). For example, inhibitory G proteins, $G_{\alpha i1}$ and $G_{\alpha i2}$, and $G_{\alpha o}$ levels are reduced (32 to 49%) in the NAC of METH (and heroin) abusers.[233] These results are consistent with rodent studies showing that cocaine and heroin decrease inhibitory G protein levels in the NAC.[185,186] Experiments exploring the effects of METH specifically on G protein levels have not been conducted in animals. Although the lower inhibitory G protein levels could represent a preexisting deficit, it is more likely that they are the result of neural adaptations employed to restore balance in response to chronic METH stimulation. Inhibitory G proteins are linked to a number of receptors including D_2. A compensatory down-regulation of the D_2 receptor pathway, or D_2 receptors specifically, is consistent with imaging studies showing that this receptor is significantly decreased in METH abusers.[234] Moreover, D_1 receptor protein is significantly increased, whereas D_2 receptors are marginally decreased in the NAC of METH abusers.[235] The increase in D_1 receptors could also be part of the compensatory homeostatic mechanism engendered to oppose overstimulation of the D_2 pathway. Indeed, this scenario is supported by findings in a number of animal studies.[179,185,191] Yet, although total D_1 receptor protein is increased, evidence shows that DA stimulation of adenylyl cyclase activity is decreased by 25 to 30% in the NAC of human METH abusers.[236] These results call for caution in predicting functional abnormalities based on receptor and G protein concentrations. Additional studies in human brain are needed to further characterize intracellular neuroadaptive changes in the addicted state.

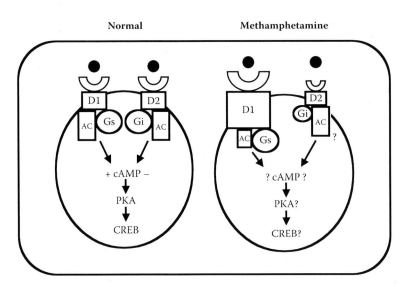

Figure 4.1   Hypothetical schematic diagram depicting cellular changes in the NAC based on post-mortem analysis of human brain. Normally, DA stimulates D_1 receptors coupled to the G_s G protein that stimulates the formation of cAMP via adenylyl cyclase (AC). The D_2 receptor is coupled to the inhibitory $G_{\alpha i}$ G protein that inhibits the formation of cAMP by AC. Accumulation of cAMP frees the catalytic subunits of cAMP-dependent protein kinase (PKA) to enter the nucleus and phosphorylate CREB. Most all drugs of abuse alter this cascade to an extent. Although evidence is limited, methamphetamine abusers show increased D_1 receptor protein levels and decreased DA-stimulated AC (indicated by the increased and reduced size, respectively). D_2 receptor protein levels are marginally decreased and $G_{\alpha i}$ and $G_{\alpha o}$ G proteins are significantly decreased in METH abusers. These changes may represent adaptations aimed at regaining homeostasis. Decreases in D_2 receptors and inhibitory G proteins may compensate for overstimulation with METH-induced supraphysiological levels of extracellular DA. Increased D_1 receptor levels could also be considered compensatory for the cellular effects rendered through overstimulated D_2 receptors. However, D_1-stimulated AC activity is impaired in human METH abusers. These results indicate possible downregulation or tolerance in both pathways by different mechanisms.

4.5 NEUROTOXICITY ASSOCIATED WITH METH CONSUMPTION

Evidence from rodent, nonhuman primate, and post-mortem human studies indicates that METH is highly toxic to the CNS. This section briefly reviews neurotoxicity associated with METH abuse with particular attention on monoamines. Excellent and detailed reviews have been published elsewhere.[77,237-242]

METH-evoked neurotoxicity in the striatal DA system has been characterized in a number of species. For example, acute and chronic administration leads to striatal DA depletion, damaged nerve terminals,[243-246] and altered DAT,[64,243,247,248] tyrosine hydroxylase,[249,250] and VMAT.[76,251,252] The hyperthermic-inducing effects of METH play a role in toxicity. Experiments in rats show that blocking METH-facilitated increases in body temperature[66] is neuroprotective[253,254] perhaps by decreasing damage caused by reactive oxygen species formed from supraphysiological levels of extracellular DA.[66] Further, evidence indicates that terminal regions of the nigrostriatal DA system are especially susceptible to the toxic effects of METH,[255] whereas the VTA–NAC reward pathway is less affected.[256] Similar to DA, acute and chronic METH exposure decreases tryptophan hydroxylase, SERT, and depletes 5-HT.[82,251]

Analogous to most rodent studies, nonhuman primate studies show METH-induced deficits in DAT, VMAT, and DA.[256-258] Interestingly, long-term experiments indicate that some of these effects reverse over time,[259,260] particularly when METH dosage regimen resembles the "binge"-like intake patterns seen in humans. Correspondingly, rodent and primate studies suggest that metabolite parameters in DA and 5-HT neurotransmitter systems and behavior appear to normalize over time; however, the extent of recovery depends on dose and length of drug exposure.[259,261–264]

Post-mortem human studies *partially* confirm preclinical findings in animals. METH abusers have decreased striatal tyrosine hydroxylase, DA, and DAT in the NAC and striatum.[265] Yet pre-synaptic markers VMAT and DOPA decarboxylase are not altered.[266] These findings suggest that there is no permanent damage to neurons in humans and confirms the results of one study in monkeys showing that nigral cell bodies are preserved following recovery.[255] However, evidence from imaging studies indicate, no matter the length of time of recovery, deficits remain[267] (see below). A number of factors, however, may explain these discrepant results between animal and post-mortem studies such as dose, duration of abuse, young vs. old population, and past drug histories. Taken together, post-mortem evidence supports that the human brain is susceptible to METH-induced alterations in DAergic parameters.

4.5.1   Oxidative Stress: A Possible Cause of METH-Induced Neurotoxicity

The exact mechanisms responsible for METH-induced neurotoxicity have not been fully defined. As mentioned previously, however, a large body of work implicates oxidative stress inflicted by reactive oxygen species in damaging neurons. Although GLU plays a significant role in the destructive effects of METH,[268] it is clear that excess DA is required for neurotoxicity to occur. Reactive species can form from oxidation of DA, DA auto-oxidation, and disruption of mitochondria.[77] Pretreatment with DA-synthesis inhibitors prevents METH-induced damage in both DA and 5-HT systems and L-dopa reverses this effect.[82,269–271] METH administration also induces the formation of an endogenous neurotoxin 6-hydroxydopamine (6-OHDA), used experimentally to induce DA-specific lesions.[272,273] Further, studies in genetically modified mice have shown that the degree of damage is mediated, in part, by a number of enzymes. Mice over-expressing the reducing enzyme superoxide dismutase (SOD) show reduced METH-induced neurotoxicity.[274–277] In contrast, mice devoid of the reactive-species-producing enzyme nitrous oxide synthase are resistant to the toxic effects of METH.[278] METH abnormally redistributes DA into the oxidizing environment of the neuron's cytoplasm from the reducing environment of the synaptic vesicles leading to possible damage to the neuron. Support for this assumption comes from experiments in which mice lacking the VMAT-2, which sequesters DA into synaptic vesicles, show exacerbated METH-induced damage in the DA system.[279] Also consistent with this notion, antioxidants including ascorbic acid,[280] vitamin E,[274] nicotinamide,[281] melatonin,[282,283] and selenium[284–286] attenuate METH-induced neurotoxicity.

4.5.2   METH-Induced Effects in Human Brain: Imaging Studies

Advances in imaging techniques have furthered our knowledge of the neural circuits involved in addiction. Positron emission tomography (PET), single photon emission tomography (SPECT), and function magnetic resonance imaging (fMRI), among others, allow measurement of relevant neuropharmacological parameters in the living brain. Recent studies using these techniques in METH abusers reveal a number of abnormalities. For example, a PET [^{18}F]fluorodeoxyglucose (a marker of brain glucose metabolism) study in detoxified METH abusers showed hypermetabolism in the parietal cortex and hypometabolism in the striatum (caudate and putamen), suggesting a dysregulation between DAergic and non-DAergic mechanisms.[267] Compared to controls, current METH abusers undergoing a vigilance task exhibit lower metabolism in areas of the brain implicated in mood (anterior cingulate, insula and orbitofrontal area, middle and posterior cingulate, amygdala, ventral striatum, and cerebellum) as also measured by PET [^{18}F]fluorodeoxyglucose.[287] Two additional SPECT ^{99m}Tc-hexamethylpropylene-amine-oxime (HMPAO) studies corroborated with PET results show abnormal perfusion profiles in METH abusers.[288,289] Greater thalamic but not striatal (caudate and NAC) metabolism is apparent in METH abusers abstinent <6 months compared to 12 to 17 months.[290] The persistent decrease in striatal metabolism suggests long-lasting changes. Consistent with findings in post-mortem human brain, individuals with METH abuse histories have significant decreases in binding of the PET DAT radioligand [^{11}C]WIN-35,428 in the caudate and putamen.[291] Other PET studies using compounds targeting the DAT have confirmed decreased DAT binding in the caudate and putamen in current abusers,[292] recently detoxified subjects,[293] and in those abstinent for upwards of 11 months.[234] Studies have also linked abnormalities in D$_2$ receptors and substance abuse given that lower levels are found in alcoholics,[294] and cocaine[295] and heroin[296] abusers. In line with these data, D$_2$ receptor levels are lower in the caudate and putamen of METH abusers, as measured by PET [^{11}C]raclopride[267] (but see Reference 293). Taken together, these data clearly show abnormal brain function in METH abusers. Future studies are needed to determine permanent neurochemical deficits through longitudinal studies and possible therapies to reverse these deficits.

4.6 CONCLUSIONS

METH continues to be a major public health concern in the U.S. and other parts of the world. The dramatic increase in METH patient admissions and untoward effects on social, community, and familial sectors underscores the need for effective treatments. Yet despite significant advances in our understanding of the neurobiological mechanisms that govern its addictive properties, effective pharmacotherapies have not emerged.[297,298] Long-term social support and pharmacological treatments aimed at relieving the protracted anhedonia, dysphoria, anxiety, severe craving from METH withdrawal, and maladaptive drug-seeking are badly needed. Campaigns to alert the health care community of the importance in identification and proper treatment of the METH-addicted patient are paramount. It is hoped that further studies on the neurobiology of METH bring a clearer understanding of the mechanisms that underlie its deleterious effects.

REFERENCES

1. NIDA. NIDA Research Report: Methamphetamine: Abuse and Addiction, NIDA Research Report, Vol. 2005, 2002.
2. DEA. Total of All Meth Clandestine Laboratory Incidents, Calendar Year 2003, Vol. 2005, Domestic Strategic Intelligence Unit (NDAS) of the Office of Domestic Intelligence, 2003.
3. DEA. Drug Trafficking in the United States, Vol. 2005, Domestic Strategic Intelligence Unit (NDAS) of the Office of Domestic Intelligence, 2001.
4. Alles, G.A. The comparative physiological actions of the DL-beta-phenylisopropylamines. I. Pressor effect and toxicity. *J. Pharmacol. Exp. Ther.* 47:339, 1933.
5. Miller, M.A., Hughes, A.L. Epidemiology of amphetamine use in the United States, in Cho, A.K., Segal, D.S., Eds., *Amphetamine and Its Analogs.* Academic Press, San Diego, 1994, 503.
6. Caldwell, J.A., Caldwell, J.L., Darlington, K.K. Utility of dextroamphetamine for attenuating the impact of sleep deprivation in pilots. *Aviat. Space Environ. Med.* 74:1125, 2003.
7. Fukui, S., Wada, K., Iyo, M. Epidemiology of amphetamine abuse in Japan and its social implications, in Cho, A.K., Segal, D.S., Eds., *Amphetamine and Its Analogs.* Academic Press, San Diego, 1994, 503.
8. Anglin, M.D., Burke, C., Perrochet, B., Stamper, E., Dawud-Noursi, S. History of the methamphetamine problem. *J. Psychoactive Drugs* 32:137, 2000.
9. Ujike, H., Mitsumoto, S. Clinical features of sensitization to methamphetamine observed in patients with methamphetamine dependence and psychosis. *Ann. N.Y. Acad. Sci.* 1025:279, 2004.
10. Thompson, H.S. *Hell's Angels.* Ballantine Books, New York, 1966.
11. Lake, C.R., Quirk, R.S. CNS stimulants and the look-alike drugs. *Psychiatr. Clin. North Am.* 7:689, 1984.
12. Sulzer, D., Sonders, M.S., Poulsen, N.W., Galli, A. Mechanisms of neurotransmitter release by amphetamines: a review. *Prog. Neurobiol.* 75:406, 2005.

13. Cook, C.E., Jeffcoat, A.R., Hill, J.M. et al. Pharmacokinetics of methamphetamine self-administered to human subjects by smoking S-(+)-methamphetamine hydrochloride. *Drug Metab. Dispos.* 21:717, 1993.

14. Cook, C.E., Jeffcoat, A.R., Sadler, B.M. et al. Pharmacokinetics of oral methamphetamine and effects of repeated daily dosing in humans. *Drug Metab. Dispos.* 20:856, 1992.

15. Cho, A.K., Melega, W.P., Kuczenski, R., Segal, D.S. Relevance of pharmacokinetic parameters in animal models of methamphetamine abuse. *Synapse.* 39:161, 2001.

16. Fischman, M.W., Schuster, C.R. Tolerance development to chronic methamphetamine intoxication in the rhesus monkey. *Pharmacol. Biochem. Behav.* 2:503, 1974.

17. Perez-Reyes, M., White, W.R., McDonald, S.A. et al. Clinical effects of daily methamphetamine administration. *Clin. Neuropharmacol.* 14:352, 1991.

18. Martin, W.R., Sloan, J.W., Sapira, J.D., Jasinski, D.R. Physiologic, subjective, and behavioral effects of amphetamine, methamphetamine, ephedrine, phenmetrazine, and methylphenidate in man. *Clin. Pharmacol. Ther.* 12:245, 1971.

19. Gawin, F.H., Ellinwood, E.H.J. Cocaine and other stimulants. Actions, abuse, and treatment. *N. Engl. J. Med.* 318:1173, 1988.

20. Cretzmeyer, M., Sarrazin, M.V., Huber, D.L., Block, R.I., Hall, J.A. Treatment of methamphetamine abuse: research findings and clinical directions. *J. Subst. Abuse Treat.* 24:267, 2003.

21. Newton, T.F., Kalechstein, A.D., Duran, S., Vansluis, N., Ling, W. Methamphetamine abstinence syndrome: preliminary findings. *Am. J. Addict.* 13:248, 2004.

22. Rawson, R.A., Huber, A., Brethen, P. et al. Status of methamphetamine users 2–5 years after outpatient treatment. *J. Addict. Dis.* 21:107, 2002.

23. Cantwell, B., McBride, A.J. Self-detoxification by amphetamine-dependent patients: a pilot study. *Drug Alcohol Depend.* 49:157, 1998.

24. Weddington, W.W., Brown, B.S., Haertzen, C.A. et al. Changes in mood, craving, and sleep during short-term abstinence reported by male cocaine addicts. A controlled, residential study. *Arch. Gen. Psychiatry.* 47:861, 1990.

25. Yui, K., Goto, K., Ikemoto, S. et al. Neurobiological basis of relapse prediction in stimulant-induced psychosis and schizophrenia: the role of sensitization. *Mol. Psychiatry.* 4:512, 1999.

26. Srisurapanont, M., Ali, R., Marsden, J. et al. Psychotic symptoms in methamphetamine psychotic in-patients. *Int. J. Neuropsychopharmacol.* 6:347, 2003.

27. Yui, K., Ikemoto, S., Ishiguro, T., Goto, K. Studies of amphetamine or methamphetamine psychosis in Japan: relation of methamphetamine psychosis to schizophrenia. *Ann. N.Y. Acad. Sci.* 914:1, 2000.

28. Akiyama, K., Kanzaki, A., Tsuchida, K., Ujike, H. Methamphetamine-induced behavioral sensitization and its implications for relapse of schizophrenia. *Schizophrenia Res.* 12:251, 1994.

29. Heffner, T.G., Hartman, J.A., Seiden, L.S. Feeding increases dopamine metabolism in the rat brain. *Science.* 208:1168, 1980.

30. Wise, R.A., Rompre, P.P. Brain dopamine and reward, in Rosenweig, M., Porter, L., Eds. *Annual Review of Psychology.* Annual Review, Inc., Palo Alto, CA, 1989, 191.

31. Roberts, D.C., Koob, G.F. Disruption of cocaine self-administration following 6-hydroxydopamine lesions of the ventral tegmental. *Pharmacol. Biochem. Behav.* 17:901, 1982.

32. DiChiara, G., Imperato, A. Drugs abused by humans preferentially increase synaptic dopamine concentrations in the mesolimbic system of freely moving rats. *Proc. Natl. Acad. Sci. U.S.A.* 85:5274, 1988.

33. Robinson, T.E., Berridge, K.C. Addiction. *Annu. Rev. Psychol.* 54:75, 2003.

34. Hart, C.L., Ward, A.S., Haney, M., Foltin, R.W., Fischman, M.W. Methamphetamine self-administration by humans. *Psychopharmacology.* 157:75, 2001.

35. Fischman, M.W., Schuster, C.R. Cocaine self-administration in humans. *Fed. Proc.* 41:241, 1982.

36. Vollm, B.A., deAraujo, I.E., Cowen, P.J. et al. Methamphetamine activates reward circuitry in drug naive human subjects. *Neuropsychopharmacology.* 29:1715, 2004.

37. Pettit, H.O., Ettenberg, A., Bloom, F.E., Koob, G.F. Destruction of dopamine in the nucleus accumbens selectively attenuates cocaine but not heroin self-administration in rats. *Psychopharmacology.* 84:167, 1984.

38. Zito, K.A., Vickers, G., Roberts, D.C. Disruption of cocaine and heroin self-administration following kainic acid lesions of the nucleus accumbens. *Pharmacol. Biochem. Behav.* 23:1029, 1985.

39. Glowinski, J., Iversen, L.L. Regional studies of catecholamines in the rat brain. I. The disposition of [³H]norepinephrine, [³H]dopamine and [³H]dopa in various regions of the brain. *J. Neurochem.* 13:655, 1966.

40. Snyder, S.H., Coyle, J.T. Regional differences in ³H-norepinephrine and ³H-dopamine uptake into rat brain homogenates. *J. Pharmacol. Exp. Ther.* 165:78, 1969.

41. Robinson, T.E., Becker, J.B. Enduring changes in brain and behavior produced by chronic amphetamine administration: a review and evaluation of animal models of amphetamine psychosis. *Brain Res. Rev.* 11:157, 1986.

42. Robinson, T.E., Berridge, K.C. The neural basis of drug craving: an incentive-sensitization theory of addiction. *Brain Res. Rev.* 18:247, 1993.

43. Ellinwood, E.H., Sudilovski, A., Nelson, L.J. Evolving behavior in the clinical and experimental amphetamine (model) psychosis. *Am. J. Psychiatry.* 34:1088, 1974.

44. Pierce, R.C., Kalivas, P.W. A circuitry model of the expression of behavioral sensitization to amphetamine-like psychostimulants. *Brain Res. Rev.* 25:192, 1997.

45. Koob, G.F., LeMoal, M. Drug addiction, dysregulation of reward, and allostasis. *Neuropsychopharmacology.* 24:97, 2001.

46. Ritz, M.C., Lamb, R.J., Goldberg, S.R., Kuhar, M.J. Cocaine receptors on dopamine transporters are related to self-administration of cocaine. *Science.* 237:1219, 1987.

47. Sulzer, D., Sonders, M.S., Poulsen, N.W., Galli, A. Mechanisms of neurotransmitter release by amphetamines: a review. *Prog. Neurobiol.* 75:406, 2005.

48. Liang, N.Y., Rutledge, C.O. Comparison of the release of [3H]dopamine from isolated corpus striatum by amphetamine, fenfluramine and unlabelled dopamine. *Biochem. Pharmacol.* 31:983, 1982.

49. Giros, B., Jabar, M., Jones, S.R., Wightman, R.M., Caron, M.G. Hyperlocomotion and indifference to cocaine and amphetamine in mice lacking the dopamine transporter. *Nature.* 379:606, 1996.

50. Kashihara, K., Hamamura, T., Okumura, K., Otsuki, S. Methamphetamine-induced dopamine release in the medial frontal cortex of freely moving rats. *Jpn. J. Psychiatry Neurol.* 45:677, 1991.

51. Melega, W.P., Williams, A.E., Schumitz, D.A., DiStefano, E.W., Cho, A.K. Pharmacokinetic and pharmacodynamic analysis of the actions of D-amphetamine and D-methamphetamine on the dopamine terminal. *J. Pharmacol. Exp. Ther.* 274:90, 1995.

52. Baumann, M.H., Ayestas, M.A., Sharpe, L.G., et al. Persistent antagonism of methamphetamine-induced dopamine release in rats pretreated with GBR12909 decanoate. *J. Pharmacol. Exp. Ther.* 301:1190, 2002.

53. Fischer, J., Cho, A. Chemical release of dopamine from striatal homogenates: evidence for an exchange diffusion model. *J. Pharmacol. Exp. Ther.* 208:203, 1979.

54. Jones, S.R., Gainetdinov, R.R., Wightman, R.M., Caron, M.G. Mechanisms of amphetamine action revealed in mice lacking the dopamine transporter. *J. Neurosci.* 18:1979, 1998.

55. Cubells, J.F., Rayport, S., Raygendran, G., Sulzer, D. Methamphetamine neurotoxicity involves vacuolation of endocytic organelles and dopamine-dependent intracellular oxidative stress. *J. Neurosci.* 14:2260, 1994.

56. Uramura, K., Yada, T., Muroya, S. et al. Methamphetamine induces cytosolic Ca^{2+} oscillations in the VTA dopamine neurons. *NeuroReport* 11:1057, 2000.

57. Amano, T., Maztsubayashi, H., Seki, T., Sasa, M., Sakai, N. Repeated administration of methamphetamine causes hypersensitivity of D_2 receptor in rat ventral tegmental area. *Neurosci. Lett.* 347:89, 2003.

58. Higashi, H., Inanaga, K., Nishi, S., Uchimura, N. Enhancement of dopamine actions on rat nucleus accumbens neurones *in vitro* after methamphetamine pre-treatment. *J. Physiol.* 408:587, 1989.

59. Chiueh, C.C., Miyake, J., Peng, M.T. Role of dopamine autoxidation, hydroxyl radical generation, and calcium overload in underlying mechanisms involved in MPTP-induced parkinsonism. *Adv. Neurol.* 60:251, 1993.

60. Berman, S.B., Zigmond, M.J., Hastings, T.G. Modification of dopamine transporter function: effect of reactive oxygen species and dopamine. *J. Neurochem.* 67:593, 1996.

61. Fleckenstein, A.E., Metzger, R.R., Beyeler, M.L., Gibb, J.W., Hanson, G.R. Oxygen radicals diminish dopamine transporter function in rat striatum. *Eur. J. Pharmacol.* 334:111, 1997.

62. Fornstedt, B., Carlsson, A. A marked rise in 5-S-cysteinyl-dopamine levels in guinea-pig striatum following reserpine treatment. *J. Neural Transm.* 76:155, 1989.

63. LaVoie, M.J., Hastings, T.G. Dopamine quinone formation and protein modification associated with the striatal neurotoxicity of methamphetamine: evidence against a role for extracellular dopamine. *J. Neurosci.* 19:1484, 1999.

64. Fleckenstein, A.E., Metzger, R.R., Wilkins, D.G., Gibb, J.W., Hanson, G.R. Rapid and reversible effects of methamphetamine on dopamine transporters. *J. Pharmacol. Exp. Ther.* 282:834, 1997.

65. Kokoshka, J.M., Vaughan, R.A., Hanson, G.R., Fleckenstein, A.E. Nature of methamphetamine-induced rapid and reversible changes in dopamine transporters. *Eur. J. Pharmacol.* 361:269, 1998.

66. Metzger, R.R., Haughey, H.M., Wilkins, D.G. et al. Methamphetamine-induced rapid decrease in dopamine transporter function: role of dopamine and hyperthermia. *J. Pharmacol. Exp. Ther.* 295:1077, 2000.

67. Broening, H.W., Morford, L.L., Vorhees, C.V. Interactions of dopamine D_1 and D_2 receptor antagonists with D-methamphetamine-induced hyperthermia and striatal dopamine and serotonin reductions. *Synapse.* 56:84, 2005.

68. Hanson, G.R., Gibb, J.W., Metzger, R.R., Kokoshka, J.M., Fleckenstein, A.E. Methamphetamine-induced rapid and reversible reduction in the activities of tryptophan hydroxylase and dopamine transporters: oxidative consequences? *Ann. N.Y. Acad. Sci.* 844:103, 1998.

69. Giovanni, A., Liang, L.P., Hastings, T.G., Zigmond, M.J. Estimating hydroxyl radical content in rat brain using systemic and intraventricular salicylate: impact of methamphetamine. *J. Neurochem.* 64:1819, 1995.

70. Liu, Y., Edwards, R.H. The role of vesicular transport proteins in synaptic transmission and neural degeneration. *Annu. Rev. Neurosci.* 20:125, 1997.

71. Brown, J.M., Hanson, G.R., Fleckenstein, A.E. Methamphetamine rapidly decreases vesicular dopamine uptake. *J. Neurochem.* 74:2221, 2000.

72. Hogan, K.A., Staal, R.G., Sonsalla, P.K. Analysis of VMAT2 binding after methamphetamine or MPTP treatment: disparity between homogenates. *J. Neurochem.* 74:2217, 2000.

73. Brown, J.M., Riddle, E.L., Sandoval, V. et al. A single methamphetamine administration rapidly decreases vesicular dopamine uptake. *J. Pharmacol. Exp. Ther.* 302:497, 2002.

74. Riddle, E.L., Topham, M.K., Haycock, J.W., Hanson, G.R., Fleckenstein, A.E. Differential trafficking of the vesicular monoamine transporter-2 by methamphetamine and cocaine. *Eur. J. Pharmacol.* 449:71, 2002.

75. Eyerman, D.J., Yamamoto, B.K. Lobeline attenuates methamphetamine-induced changes in vesicular monoamine transporter 2 immunoreactivity and monoamine depletions in the striatum. *J. Pharmacol. Exp. Ther.* 312:160, 2005.

76. Fleckenstein, A.E., Hanson, G.R. Impact of psychostimulants on vesicular monoamine transporter function. *Eur. J. Pharmacol.* 479:283, 2003.

77. Davidson, C., Gow, A.J., Lee, T.H., Ellinwood, E.H. Methamphetamine neurotoxicity: necrotic and apoptotic mechanisms and relevance to human abuse and treatment. *Brain Res. Rev.* 36:1, 2001.

78. Rocher, C., Gardier, A.M. Effects of repeated systemic administration of *d*-fenfluramine on serotonin and glutamate release in rat ventral hippocampus: comparison with methamphetamine using *in vivo* microdialysis. *Naunyn Schmiedeberg's Arch. Pharmacol.* 363:422, 2001.

79. Bakhit, C., Morgan, M.E., Peat, M.A., Gibb, J.W. Long-term effects of methamphetamine on the synthesis and metabolism of 5-hydroxytryptamine in various regions of the rat brain. *Neuropharmacology.* 20:1135, 1981.

80. Morgan, M.E., Gibb, J.W. Short-term and long-term effects of methamphetamine on biogenic amine metabolism in extra-striatal dopaminergic nuclei. *Neuropharmacology.* 19:989, 1980.

81. Haughey, H.M., Fleckenstein, A.E., Hanson, G.R. Differential regional effects of methamphetamine on the activities of tryptophan and tyrosine hydroxylase. *J. Neurochem.* 72:661, 1999.

82. Hotchkiss, A.J., Morgan, M.E., Gibb, J.W. The long-term effects of multiple doses of methamphetamine on neostriatal tryptophan hydroxylase, tyrosine hydroxylase, choline acetyltransferase and glutamate decarboxylase activities. *Life Sci.* 25:1373, 1979.

83. Hanson, G.R., Rau, K.S., Fleckenstein, A.E. The methamphetamine experience: a NIDA partnership. *Neuropharmacology.* 47:92, 2004.

84. Kokoshka, J.M., Metzger, R.R., Wilkins, D.G. et al. Methamphetamine treatment rapidly inhibits serotonin, but not glutamate, transporters in rat brain. *Brain Res.* 799:78, 1998.

85. Fleckenstein, A.E., Haughey, H.M., Metzger, R.R. et al. Differential effects of psychostimulants and related agents on dopaminergic and serotonergic transporter function. *Eur. J. Pharmacol.* 382:45, 1999.

86. Kovachich, G.B., Aronson, C.E., Brunswick, D.J. Effects of high-dose methamphetamine administration on serotonin uptake sites in rat brain measured using [3H]cyanoimipramine autoradiography. *Brain Res.* 505:123, 1989.

87. Wrona, M.Z., Dryhurst, G. Oxidation of serotonin by superoxide radical: implications to neurodegenerative brain disorders. *Chem. Res. Toxicol.* 11:639, 1998.

88. Haughey, H.M., Fleckenstein, A.E., Metzger, R.R., Hanson, G.R. The effects of methamphetamine on serotonin transporter activity: role of dopamine and hyperthermia. *J. Neurochem.* 75:1608, 2000.

89. Schmidt, C.J., Gibb, J.W. Role of the serotonin uptake carrier in the neurochemical response to methamphetamine: effects of citalopram and chlorimipramine. *Neurochem. Res.* 10:637, 1985.

90. Kuczenski, R., Segal, D.S., Cho, A.K., Melega, W. Hippocampus norepinephrine, caudate dopamine and serotonin, and behavioral responses to the stereoisomers of amphetamine and methamphetamine. *J. Neurosci.* 15:1308, 1995.

91. Fornai, F., Alessandri, M.G., Torracca, M.T. et al. Noradrenergic modulation of methamphetamine-induced striatal dopamine depletion. *Ann. N.Y. Acad. Sci.* 844:166, 1998.

92. Bakhit, C., Morgan, M.E., Gibb, J.W. Propranolol differentially blocks the methamphetamine-induced depression of tryptophan hydroxylase in various rat brain regions. *Neurosci. Lett.* 99, 1981.

93. Haughey, H.M., Brown, J.M., Wilkins, D.G., Hanson, G.R., Fleckenstein, A.E. Differential effects of methamphetamine on Na(+)/Cl(−)-dependent transporters. *Brain Res.* 863:59, 2000.

94. Preseton, K.L., Wagner, G.C., Schuster, C.R., Seiden, L.S. Long-term effects of repeated methylamphetamine administration on monoamine neurons in the rhesus monkey brain. *Brain Res.* 338:243, 1985.

95. Kitanaka, N., Kitanaka, J., Takemura, M. Behavioral sensitization and alteration in monoamine metabolism in mice after single versus repeated methamphetamine administration. *Eur. J. Pharmacol.* 474:63, 2003.

96. Rothman, R.B., Baumann, M.H., Dersch, C.M. et al. Amphetamine-type central nervous system stimulants release norepinephrine more potently than they release dopamine and serotonin. *Synapse.* 39:32, 2001.

97. Zhang, X.Y., Kosten, T.A. Prazosin, an alpha-1 adrenergic antagonist, reduces cocaine-induced reinstatement of drug-seeking. *Biol. Psychiatry.* 57:1202, 2005.

98. Yui, K., Goto, K., Ikemoto, S. The role of noradrenergic and dopaminergic hyperactivity in the development of spontaneous recurrence of methamphetamine psychosis and susceptibility to episode recurrence. *Ann. N.Y. Acad. Sci.* 1025:296, 2004.

99. Wise, R.A. Dopamine, learning and motivation. *Nat. Rev. Neurosci.* 5:483, 2004.

100. Robbins, T.W., Everitt, B.J. Limbic-striatal memory systems and drug addiction. *Neurobiol. Learn. Mem.* 78:625, 2002.

101. Jones, S., Bonci, A. Synaptic plasticity and drug addiction. *Curr. Opin. Pharmacol.* 5:20, 2005.

102. Zetterstrom, T., Sharp, T., Collin, A.K., Ungerstedt, U. *In vivo* measurement of extracellular dopamine and DOPAC in rat striatum after various dopamine-releasing drugs; implications for the origin of extracellular DOPAC. *Eur. J. Pharmacol.* 148:327, 1988.

103. Shoblock, J.R., Sullivan, E.B., Maisonneuve, I.M., Glick, S.D. Neurochemical and behavioral differences between d-methamphetamine and d-amphetamine in rats. *Psychopharmacology.* 165:359, 2003.

104. Ujike, H., Onoue, T., Akiyama, K., Hamamura, T., Otsuki, S. Effects of selective D-1 and D-2 dopamine antagonists on development of methamphetamine-induced behavioral sensitization. *Psychopharmacology.* 98:89, 1989.

104a. Witkin, J.M., Savtchenko, N., Mashkovsky, M., Beekman, M., Munzar, P., Gasior, M., Goldberg, S.T., Ungard, J.T., Kim, J., Shippenberg, T., Chefer, V. Behavioral, toxic, and neurochemical effects of sydnocarb, a novel psychomotor stimulant: comparisons with methamphetamine. *J. Pharmacol. Exp. Ther.* 288:1298, 1999.

105. Post, R.M. Intermittent versus continuous stimulation: effect of time interval on the development of sensitization. *Life Sci.* 26:1275, 1980.

106. Davidson, C., Lee, T.H., Ellinwood, E.H. Acute and chronic continuous methamphetamine have different long-term behavioral and neurochemical consequences. *Neurochem. Int.* 46:189, 2005.

107. Yokel, R.A., Pickens, R. Self-administration of optical isomers of amphetamine and methylamphet-amine by rats. *J. Pharmacol. Exp. Ther.* 187:27, 1973.

108. Balster, R.L., Schuster, C.R. A comparison of *d*-amphetamine, *l*-amphetamine, and methamphetamine self-administration in rhesus monkeys. *Pharmacol. Biochem. Behav.* 1:67, 1973.

109. Balster, R.L., Kilbey, M.M., Ellinwood, E.H.J. Methamphetamine self-administration in the cat. *Psychopharmacologia.* 46:229, 1976.

110. Woolverton, W.L., Cervo, L., Johanson, C.E. Effects of repeated methamphetamine administration on methamphetamine self-administration in rhesus monkeys. *Pharmacol. Biochem. Behav.* 21:737, 1984.

111. Stefanski, R., Ladenheim, B., Lee, S.H., Cadet, J.L., Goldberg, S.R. Neuroadaptations in the dopam-inergic system after active self-administration but not after passive administration of methamphet-amine. *Eur. J. Pharmacol.* 371:123, 1999.

112. Stefanski, R., Lee, S.H., Yasar, S., Cadet, J.L., Goldberg, S.R. Lack of persistent changes in the dopaminergic system of rats withdrawn from methamphetamine self-administration. *Eur. J. Pharma-col.* 439:59, 2002.

113. Ranaldi, R., Wise, R.A. Intravenous self-administration of methamphetamine-heroin (speedball) combinations under a progressive-ratio schedule of reinforcement in rats. *Neuroreport.* 11:2621, 2000.

114. Wong, A.H.C., VanTol, H.H.M. The dopamine D_4 receptors and mechanisms of antipsychotic atypi-cality. *Prog. Neuropsychopharmacol. Biol. Psychiatry.* 27:1091, 2003.

115. Okuyama, S., Kawashima, N., Chaki, S. et al. A selective dopamine D_4 receptor antagonist, NRA0160: a preclinical neuropharmacological profile. *Life Sci.* 65:2109, 1999.

116. Rothman, R.B., Partilla, J.S., Baumann, M.H. et al. Neurochemical neutralization of methamphetamine with high-affinity nonselective inhibitors of biogenic amine transporters: a pharmacological strategy for treating stimulant abuse. *Synapse.* 35:222, 2000.

117. Schenk, S. Effects of GBR 12909, WIN 35,428 and indatraline on cocaine self-administration and cocaine seeking in rats. *Psychopharmacology.* 160:263, 2002.

118. Ranaldi, R., Anderson, K.G., Carroll, F.I., Woolverton, W.L. Reinforcing and discriminative stimulus effects of RTI 111, a 3-phenyltropane analog, in rhesus monkeys: interaction with methamphetamine. *Psychopharmacology.* 53:103, 2000.

119. Carroll, K.M., Fenton, L.R., Ball, S.A. et al. Efficacy of disulfiram and cognitive behavior therapy in cocaine-dependent outpatients: a randomized placebo-controlled trial. *Arch. Gen. Psychiatry.* 61:264, 2004.

120. Zaczek, R., Culp, S., DeSouza, E.B. Interactions of [3H]amphetamine with rat brain synaptosomes. II. Active transport. *J. Pharmacol. Exp. Ther.* 257:830, 1991.

121. Holtzman, S.G. Differential interaction of GBR 12909, a dopamine uptake inhibitor, with cocaine and methamphetamine in rats discriminating cocaine. *Psychopharmacology.* 155:180, 2001.

122. Rocha, B.A., Scearce-Levie, K., Lucas, J.J. et al. Increased vulnerability to cocaine in mice lacking the serotonin-1B receptor. *Nature.* 393:175, 1998.

123. Leccese, A.P., Lyness, W.H. The effects of putative 5-hydroxytryptamine receptor active agents on D-amphetamine self-administration in controls and rats with 5,7-dihydroxytryptamine median fore-brain bundle lesions. *Brain Res.* 303:153, 1984.

124. Parsons, L.H., Justice, J.B. Perfusate serotonin increases extracellular dopamine in the nucleus accum-bens as measured by *in vivo* microdialysis. *Brain Res.* 606:195, 1993.

125. Yeghiayan, S.K., Kelley, A.E., Kula, N.S., Campbell, A., Baldessarini, R.J. Role of dopamine in behavioral effects of serotonin microinjected into rat striatum. *Pharmacol. Biochem. Behav.* 56:251, 1997.

126. Berger, U.V., Gu, X.F., Azmitia, E.C. The substituted amphetamines 3,4-methylenedioxymethamphet-amine, methamphetamine, *p*-chloroamphetamine and fenfluramine induce 5-hydroxytryptamine release via a common mechanism blocked by fluoxetine and cocaine. *Eur. J. Pharmacol.* 215:153, 1992.

127. Munzar, P., Laufert, M.D., Kutkat, S.W., Novakova, J., Goldberg, S.R. Effects of various serotonin agonists, antagonists, and uptake inhibitors on the discriminative stimulus effects of methamphetamine in rats. *J. Pharmacol. Exp. Ther.* 291:239, 1999.

128. Ginawi, O.T., Al-Majed, A.A., Al-Suwailem, A.K., El-Hadiyah, T.M. Involvement of some 5-HT recep-tors in methamphetamine-induced locomotor activity in mice. *J. Physiol. Pharmacol.* 55:357, 2004.

129. Ginawi, O.T., Al-Majed, A.A., Al-Suwailem, A.K. NAN-190, a possible specific antagonist for methamphetamine. *Regul. Toxicol. Pharmacol.* 41:122, 2005.

130. Czoty, P.W., Ramanathan, C.R., Mutschler, N.H., Makriyannis, A., Bergman, J. Drug discrimination in methamphetamine-trained monkeys: effects of monoamine transporter inhibitors. *J. Pharmacol. Exp. Ther.* 311:720, 2004.

131. Rockhold, R.W. Glutamatergic involvement in psychomotor stimulant action. *Prog. Drug Res.* 50:155, 1998.

132. Carlezon, W.A.J., Wise, R.A. Rewarding actions of phencyclidine and related drugs in nucleus accumbens shell and frontal cortex. *J. Neurosci.* 16:3112, 1996.

133. Carlezon, W.A.J., Boundy, V.A., Haile, C.N. et al. Sensitization to morphine induced by viral-mediated gene transfer. *Science.* 277:812, 1997.

134. Feenstra, M.G., Botterblom, M.H., van Uum, J.F. Behavioral arousal and increased dopamine efflux after blockade of NMDA-receptors in the prefrontal cortex are dependent on activation of glutamatergic neurotransmission. *Neuropharmacology.* 42:752, 2002.

135. Chiamulera, C., Epping-Jordan, M.P., Zocchi, A. et al. Reinforcing and locomotor stimulant effects of cocaine are absent in mGluR5 null mutant mice. *Nat. Neurosci.* 4:873, 2001.

136. Tessari, M., Pilla, M., Andreoli, M., Hutcheson, D.M., Heidbreder, C.A. Antagonism at metabotropic glutamate 5 receptors inhibits nicotine- and cocaine-taking behaviours and prevents nicotine-triggered relapse to nicotine-seeking. *Eur. J. Pharmacol.* 19:121, 2004.

137. Larson, J., Quach, C.N., LeDuc, B.Q. et al. Effects of an AMPA receptor modulator on methamphetamine-induced hyperactivity. *Brain Res.* 738:353, 1996.

138. Witkin, J.M. Blockade of the locomotor stimulant effects of cocaine and methamphetamine by glutamate antagonists. *Life Sci.* 53:PL405, 1993.

139. Nakagawa, T., Fujio, M., Ozawa, T., Minami, M., Satoh, M. Effect of MS-153, a glutamate transporter activator, on the conditioned rewarding effects of morphine, methamphetamine and cocaine in mice. *Behav. Brain Res.* 156:233, 2005.

140. Haile, C.N., GrandPre, T., Kosten, T.A. Chronic unpredictable stress, but not chronic predictable stress, enhances the sensitivity to the behavioral effects of cocaine in rats. *Psychopharmacology.* 154:213, 2001.

141. Carlsson, M.L. On the role of cortical glutamate in obsessive-compulsive disorder and attention-deficit hyperactivity disorder, two phenomenologically antithetical conditions. *Acta Psychiatr. Scand.* 102:401, 2000.

142. Compton, W.M., Cottler, L.B., BenAbdallah, A. et al. Substance dependence and other psychiatric disorders among drug dependent subjects: race and gender correlates. *Am. J. Addict.* 9:113, 2000.

143. Guitart, X., Codony, X., Monroy, X. Sigma receptors: biology and therapeutic potential. *Psychopharmacology.* 174:301, 2004.

144. Hayashi, T., Su, T.P. Sigma-1 receptor ligands: potential in the treatment of neuropsychiatric disorders. *CNS Drugs.* 18:269, 2004.

145. Matsumoto, R.R., Liu, Y., Lerner, M., Howard, E.W., Brackett, D.J. Sigma receptors: potential medications development target for anti-cocaine agents. *Eur. J. Pharmacol.* 469:1, 2003.

146. Sharkey, J., Glen, K.A., Wolfe, S., Kuhar, M.J. Cocaine binding at sigma receptors. *Eur. J. Pharmacol.* 49:171, 1988.

147. Maurice, T., Martin-Fardon, R., Romieu, P., Matsumoto, R.R. Sigma(1) (sigma(1)) receptor antagonists represent a new strategy against cocaine addiction and toxicity. *Neurosci. Biobehav. Rev.* 26:499, 2002.

148. Nguyen, E.C., McCracken, K.A., Liu, Y., Pouw, B., Matsumoto, R.R. Involvement of sigma (sigma) receptors in the acute actions of methamphetamine: receptor binding and behavioral studies. *Neuropharmacology.* 3:00, 2005.

149. Ujike, H., Okumuma, K., Zushi, Y., Akiyama, K., Otsuki, S. Persistent supersensitivity of s receptors develops during repeated methamphetamine treatment. *Eur. J. Pharmacol.* 211:323, 1992.

150. Itzhak, Y. Repeated methamphetamine-treatment alters brain receptors. *Eur. J. Pharmacol.* 230:243, 1993.

151. Stefanski, R., Justinova, Z., Hayashi, T. et al. Sigma1 receptor upregulation after chronic methamphetamine self-administration in rats: a study with yoked controls. *Psychopharmacology.* 175:68, 2004.

152. Takahashi, S., Miwa, T., Horikomi, K. Involvement of sigma-1 receptors in methamphetamine-induced behavioral sensitization in rats. *Neurosci. Lett.* 289:21, 2000.

153. Ujike, H., Kanzaki, A., Okumuma, K., Akiyama, K., Otsuki, S. Sigma antagonist BMY 14802 prevents methamphetamine-induced sensitization. *Life Sci.* 50:PL129, 1992.

154. Hayashi, T., Su, T.-P. Regulating ankyrin dynamics: roles of sigma-1 receptors. *Proc. Natl. Acad. Sci. U.S.A.* 98:491, 2001.

155. Hayashi, T., Su, T.-P. Sigma-1 receptors (sigma-1 binding sites) form raft-like microdomains and target lipid droplets on the endoplasmic reticulum: roles in endoplasmic reticulum lipid compartmentalization and export. *J. Pharmacol. Exp. Ther.* 306:718, 2003.

156. Robinson, T.E., Kolb, B. Structural plasticity associated with exposure to drugs of abuse. *Neuropharmacology.* 47:33, 2004.

157. Maisonneuve, I.M., Glick, S.D. Anti-addictive actions of an iboga alkaloid congener: a novel mechanism for a novel treatment. *Pharmacol. Biochem. Behav.* 75:607, 2003.

158. Szumlinski, K.K., Herrick-Davis, K., Teitler, M., Maisonneuve, I.M., Glick, S.D. Interactions between iboga agents and methamphetamine sensitization: studies of locomotion and stereotypy in rats. *Psychopharmacology.* 151:234, 2000.

159. Glick, S.D., Maisonneuve, I.M., Dickinson, H.A. 18-MC reduces methamphetamine and nicotine self-administration in rats. *Neuroreport.* 11:2013, 2000.

160. Haile, C.N., During, M.J., Jatlow, P.I., Kosten, T.R., Kosten, T.A. Disulfiram facilitates the development and expression of locomotor sensitization to cocaine in rats. *Biol. Psychiatry.* 54:915, 2003.

161. Glick, S.D., Maisonneuve, I.M., Kitchen, B.A., Fleck, M.W. Antagonism of alpha 3 beta 4 nicotinic receptors as a strategy to reduce opioid and stimulant self-administration. *Eur. J. Pharmacol.* 438:99, 2002.

162. Jun, J.H., Schindler, C.W. Dextromethorphan alters methamphetamine self-administration in the rat. *Pharmacol. Biochem. Behav.* 67:405, 2000.

163. Miller, D.K., Crooks, P.A., Teng, L. et al. Lobeline inhibits the neurochemical and behavioral effects of amphetamine. *J. Pharmacol. Exp. Ther.* 296:1023, 2001.

164. Harrod, S.B., Dwoskin, L.P., Crooks, P.A., Klebaur, J.E., Bardo, M.T. Lobeline attenuates *d*-methamphetamine self-administration in rats. *J. Pharmacol. Exp. Ther.* 298:172, 2001.

165. Dwoskin, L.P., Crooks, P.A. A novel mechanism of action and potential use for lobeline as a treatment for psychostimulant abuse. *Biochem. Pharmacol.* 63:89, 2002.

166. Lendvai, B., Sershen, H., Lajtha, A. et al. Differential mechanisms involved in the effect of nicotinic agonists DMPP and lobeline to release [3H]5-HT from rat hippocampal slices. *Neuropharmacology.* 35:1769, 1996.

167. Miller, D.K., Crooks, P.A., Zhang, G. et al. Lobeline analogs with enhanced affinity and selectivity for plasmalemma and vesicular monoamine transporters. *J. Pharmacol. Exp. Ther.* 310:1035, 2004.

168. Ranaldi, R., Poeggel, K. Baclofen decreases methamphetamine self-administration in rats. *Neuroreport.* 13:1107, 2002.

169. Vinklerova, J., Novakova, J., Sulcova, A. Inhibition of methamphetamine self-administration in rats by cannabinoid receptor antagonist AM 251. *J. Psychopharmacol.* 16:139, 2002.

170. Anggadiredja, K., Nakamichi, M., Hiranita, T. et al. Endocannabinoid system modulates relapse to methamphetamine seeking: possible mediation by the arachidonic acid cascade. *Neuropsychopharmacology.* 29:1470, 2004.

171. Munzar, P., Tanda, G., Justinova, Z., Goldberg, S.R. Histamine h3 receptor antagonists potentiate methamphetamine self-administration and methamphetamine-induced accumbal dopamine release. *Neuropsychopharmacology.* 29:705, 2004.

172. McMillan, D.E., Hardwick, W.C., Li, M. et al. Effects of murine-derived anti-methamphetamine monoclonal antibodies on (+)-methamphetamine self-administration in the rat. *J. Pharmacol. Exp. Ther.* 309:1248, 2004.

173. Bailey, C.P., Connor, M. Opioids: cellular mechanisms of tolerance and physical dependence. *Curr. Opin. Pharmacol.* 5:60, 2005.

174. Pandey, S.C. The gene transcription factor cyclic AMP-responsive element binding protein: role in positive and negative affective states of alcohol addiction. *Pharmacol. Ther.* 104:47, 2004.

175. Nestler, E.J. Molecular mechanisms of drug addiction. *Neuropharmacology.* 47:24, 2004.

176. Robinson, T.E., Kolb, B. Persistent structural modifications in nucleus accumbens and prefrontal cortex neurons produced by previous experience with amphetamine. *J. Neurosci.* 17:8491, 1997.

177. Saal, D., Dong, Y., Bonci, A., Malenka, R.C. Drugs of abuse and stress trigger a common synaptic adaptation in dopamine neurons. *Neuron.* 37:577, 2003.

178. Nestler, E.J. Historical review: Molecular and cellular mechanisms of opiate and cocaine addiction. *Trends Pharmacol. Sci.* 25:210, 2004.

179. Nestler, E.J. Molecular neurobiology of addiction. *Am. J. Addict.* 10:201, 2001.

180. Self, D.W., Nestler, E.J. Molecular mechanisms of drug reinforcement and addiction. *Annu. Rev. Neurosci.* 18:463, 1995.

181. Haile, C.N., Hiroi, N., Nestler, E.J., Kosten, T.A. Differential behavioral responses to cocaine are associated with dynamics of mesolimbic dopamine proteins in Lewis and Fischer 344 rats. *Synapse.* 41:179, 2001.

182. Beitner-Johnson, D., Nestler, E.J. Morphine and cocaine exert common chronic actions on tyrosine hydroxylase in dopaminergic brain reward regions. *J. Neurochem.* 57:344, 1991.

183. Sorg, B.A., Chen, S.Y., Kalivas, P.W. Time course of tyrosine hydroxylase expression after behavioral sensitization to cocaine. *J. Pharmacol. Exp. Ther.* 266:424, 1993.

184. Henry, D.J., White, F.J. Repeated cocaine administration causes persistent enhancement of D1 dopamine receptor sensitivity within the rat nucleus accumbens. *J. Pharmacol. Exp. Ther.* 258:882, 1991.

185. Terwilliger, R.Z., Beitner-Johnson, D., Sevarino, K.A., Crain, S.M., Nestler, E.J. A general role for adaptations in G-proteins and the cyclic AMP system in mediating the chronic actions of morphine and cocaine on neuronal function. *Brain Res.* 548:100, 1991.

186. Striplin, C.D., Kalivas, P.W. Robustness of G protein changes in cocaine sensitization shown with immunoblotting. *Synapse.* 14:10, 1993.

187. Chen, J., Nye, H.E., Kelz, M.B. et al. Regulation of delta FosB and FosB-like proteins by electroconvulsive seizure and cocaine treatments. *Mol. Pharmacol.* 48:880, 1995.

188. Hiroi, N., Brown, J.R., Haile, C.N. et al. FosB mutant mice: loss of chronic cocaine induction of Fos-related proteins and heightened sensitivity to cocaine's psychomotor and rewarding effects. *Proc. Natl. Acad. Sci. U.S.A.* 94:10397, 1997.

189. Kelz, M.B., Chen, J., Carlezon, W.A.J. et al. Expression of the transcription factor deltaFosB in the brain controls sensitivity to cocaine. *Nature.* 401:272, 1999.

190. Tolliver, B.K., Ho, L.B., Reid, M.S., Berger, S.P. Evidence for involvement of ventral tegmental area cyclic AMP systems in behavioral sensitization to psychostimulants. *J. Pharmacol. Exp. Ther.* 278:411, 1996.

191. Self, D.W., Genova, L.M., Hope, B.T. et al. Involvement of cAMP-dependent protein kinase in the nucleus accumbens in cocaine self-administration and relapse of cocaine-seeking behavior. *J. Neurosci.* 18:1848, 1998.

192. Shaywitz, A.J., Greenberg, M.E. CREB: a stimulus-induced transcription factor activated by a diverse array of extracellular signals. *Annu. Rev. Biochem.* 68:821, 1999.

193. Mayr, B., Montminy, M. Transcriptional regulation by the phosphorylation-dependent factor CREB. *Nat. Rev. Mol. Cell. Biol.* 2:599, 2001.

194. Turgeon, S.M., Pollack, A.E., Fink, J.S. Enhanced CREB phosphorylation and changes in c-Fos and FRA expression in striatum accompany amphetamine sensitization. *Brain Res.* 749:120, 1997.

195. Carlezon, W.A., Duman, R.S., Nestler, E.J. The many faces of CREB. *Trends Neurosci.* 28:436, 2005.

196. Kelley, A.E. Memory and addiction: shared neural circuitry and molecular mechanisms. *Neuron.* 44:161, 2004.

197. Miserendino, M.J., Nestler, E.J. Behavioral sensitization to cocaine: modulation by the cyclic AMP system in the nucleus accumbens. *Brain Res.* 674:299, 1995.

198. Wachtel, H. Characteristic behavioural alterations in rats induced by rolipram and other selective adenosine cyclic 3´, 5´-monophosphate phosphodiesterase inhibitors. *Psychopharmacology.* 77:309, 1982.

199. Iyo, M., Bi, Y., Hashimoto, K., Inada, T., Fukui, S. Prevention of methamphetamine-induced behavioral sensitization in rats by a cyclic AMP phosphodiesterase inhibitor, rolipram. *Eur. J. Pharmacol.* 312:163, 1996.

199a. Mori, M., Baba, J., Ichimaru, Y., Suzuki, T. Effects of rolipram, a selective inhibitor of phosphodiesterase 4, on hyperlocomotion induced by several abused drugs in mice. *Jpn. J. Pharmacol.* 83:113, 2000.

200. Greengard, P., Allen, P.B., Nairn, A.C. Beyond the dopamine receptor: the DARPP-32/protein phosphatase-1 cascade. *Neuron.* 23:435, 1999.
201. Walaas, S.I., Aswad, D.W., Greengard, P. A dopamine- and cyclic AMP-regulated phosphoprotein enriched in dopamine-innervated brain regions. *Nature.* 301:69, 1983.
202. Ouimet, C.C., daCruz, E., Silva, E.F., Greengard, P. The alpha and gamma 1 isoforms of protein phosphatase 1 are highly and specifically concentrated in dendritic spines. *Proc. Natl. Acad. Sci. U.S.A.* 92:3396, 1995.
203. Yan, Z., Hsieh-Wilson, L., Feng, J. et al. Protein phosphatase 1 modulation of neostriatal AMPA channels: regulation by DARPP-32 and spinophilin. *Nat. Neurosci.* 2:13, 1999.
204. Hiroi, N., Fienberg, A.A., Haile, C.N. et al. Neuronal and behavioural abnormalities in striatal function in DARPP-32-mutant mice. *Eur. J. Neurosci.* 11:1114, 1999.
205. Fienberg, A.A., Hiroi, N., Mermelstein, P.G. et al. DARPP-32: regulator of the efficacy of dopaminergic neurotransmission. *Science.* 281:838, 1998.
206. Snyder, G.L., Allen, P.B., Fienberg, A.A. et al. Regulation of phosphorylation of the GluR1 AMPA receptor in the neostriatum by dopamine and psychostimulants *in vivo. J. Neurosci.* 20:4480, 2000.
207. Lin, X.H., Hashimoto, T., Kitamura, N. et al. Decreased calcineurin and increased phosphothreonine-DARPP-32 in the striatum of rats behaviorally sensitized to methamphetamine. *Synapse.* 44:181, 2002.
208. Chen, P.C., Chen, J.C. Enhanced Cdk5 activity and p35 translocation in the ventral striatum of acute and chronic methamphetamine-treated rats. *Neuropsychopharmacology.* 30:538, 2005.
209. Bibb, J.A., Snyder, G.L., Nishi, A. et al. Phosphorylation of DARPP-32 by Cdk5 modulates dopamine signalling in neurons. *Nature.* 402:669, 1999.
210. McClung, C.A., Ulery, P.G., Perrotti, L.I. et al. DeltaFosB: a molecular switch for long-term adaptation in the brain. *Brain Res. Mol. Brain Res.* 132:146, 2004.
211. Ouimet, C.C., Hemmings, H.C.J., Greengard, P. ARPP-21, a cyclic AMP-regulated phosphoprotein enriched in dopamine-innervated brain regions. II. Immunocytochemical localization in rat brain. *J. Neurosci.* 9:865, 1989.
212. Caporaso, G.L., Bibb, J.A., Snyder, G.L. et al. Drugs of abuse modulate the phosphorylation of ARPP-21, a cyclic AMP-regulated phosphoprotein enriched in the basal ganglia. *Neuropharmacology.* 39:1637, 2000.
213. Akiyama, K., Suemaru, J. Effect of acute and chronic administration of methamphetamine on calcium-calmodulin dependent protein kinase II activity in the rat brain. *Ann. N.Y. Acad. Sci.* 914:263, 2000.
214. Suemaru, J., Akiyama, K., Tanabe, Y., Kuroda, S. Methamphetamine decreases calcium-calmodulin dependent protein kinase II activity in discrete rat brain regions. *Synapse.* 36:155, 2000.
215. Cheung, W.Y. Calmodulin plays a pivotal role in cellular regulation. *Science.* 207:19, 1980.
216. Ishihara, T., Akiyama, K., Kashihara, K. et al. Activator protein-1 binding activities in discrete regions of rat brain after acute and chronic administration of methamphetamine. *J. Neurochem.* 67:708, 1996.
217. Bronstein, D.M., Pennypacker, K.R., Lee, H., Hong, J.S. Methamphetamine-induced changes in AP-1 binding and dynorphin in the striatum: correlated, not causally related events? *Biol. Signals.* 317, 1996.
218. Asanuma, M., Hayashi, T., Ordonex, S.V., Ogawa, N., Cadet, J.L. Direct interactions of methamphetamine with the nucleus. *Brain Res. Mol. Brain Res.* 80:237, 2000.
219. Lee, Y.W., Son, K.W., Flora, G., et al. Methamphetamine activates DNA binding of specific redox-responsive transcription factors in mouse brain. *J. Neurosci. Res.* 70:82, 2002.
220. Muratake, T., Toyooka, K., Hayashi, S. et al. Immunohistochemical changes of the transcription regulatory factors in rat striatum after methamphetamine administration. *Ann. N.Y. Acad. Sci.* 844:21, 1998.
221. Tzschentke, T.M. Measuring reward with the conditioned place preference paradigm: a comprehensive review of drug effects, recent progress and new issues. *Prog. Neurobiol.* 56:613, 1998.
221a. Carlezon, W.A., Thome, J., Olson, V.G., Lane-Ladd, S.B., Brodkin, E.S., Hiroi, N., Duman, R.S., Neve, R.L., Nestler, E.J. Regulation of cocaine reward by CREB. *Science* 282:2272, 1998.
222. Thomas, G.M., Huganir, R.L. MAPK cascade signaling and synaptic plasticity. *Nat. Rev. Neurosci.* 5:173, 2004.
223. Sweatt, J.D. Mitogen-activated protein kinases in synaptic plasticity and memory. *Curr. Opin. Neurobiol.* 14:311, 2004.

224. Wolf, M.E., Sun, X., Mangiavacchi, S., Chao, S.Z. Psychomotor stimulants and neuronal plasticity. *Neuropharmacology*. 47:61, 2004.

225. Hashimoto, T., Kajii, Y., Nishikawa, T. Psychotomimetic-induction of tissue plasminogen activator mRNA in corticostriatal neurons in rat brain. *Eur. J. Neurosci*. 10:3387, 1998.

226. Kodama, M., Akiyama, K., Ujike, H. et al. A robust increase in expression of arc gene, an effector immediate early gene, in the rat brain after acute and chronic methamphetamine administration. *Brain Res*. 796:273, 1998.

227. Takaki, M., Ujike, H., Kodama, M. et al. Increased expression of synaptophysin and stathmin mRNAs after methamphetamine administration in rat brain. *Neuroreport*. 12:1055, 2001.

228. Takaki, M., Ujike, H., Kodama, M. et al. Two kinds of mitogen-activated protein kinase phosphatases, MKP-1 and MKP-3, are differentially activated by acute and chronic methamphetamine treatment in the rat brain. *J. Neurochem*. 79:679, 2001.

229. Ujike, H., Takaki, M., Kodama, M., Kuroda, S. Gene expression related to synaptogenesis, neuritogenesis, and MAP kinase in behavioral sensitization to psychostimulants. *Ann. N.Y. Acad. Sci*. 965:55, 2002.

230. Berhow, M.T., Hiroi, N., Nestler, E.J. Regulation of ERK (extracellular signal regulated kinase), part of the neurotrophin signal transduction cascade, in the rat mesolimbic dopamine system by chronic exposure to morphine or cocaine. *J. Neurosci*. 16:4707, 1996.

231. Pierce, R.C., Pierce-Bancroft, A.F., Prasad, B.M. Neurotrophin-3 contributes to the initiation of behavioral sensitization to cocaine by activating the Ras/Mitogen-activated protein kinase signal transduction cascade. *J. Neurosci*. 19:8685, 1999.

232. Valjent, E., Corvol, J.C., Pages, C. et al. Involvement of the extracellular signal-regulated kinase cascade for cocaine-rewarding properties. *J. Neurosci*. 20:8701, 2000.

232a. Mizoguchi, H., Yamada, K., Mizuno, M., Nitta, A., Noda, Y., Nabeshima, T. Regulations of methamphetamine reward by extracellular signal-regulated kinase 1/2/ets-like gene-1 signaling pathway via the activation of dopamine receptors. *Mol. Pharmacol*. 65:1293, 2004.

233. McLeman, E.R., Warsh, J.J., Ang, L. et al. The human nucleus accumbens is highly susceptible to G protein down-regulation by methamphetamine and heroin. *J. Neurochem*. 74:2120, 2000.

234. Volkow, N.D., Chang, L., Wang, G.J. et al. Association of dopamine transporter reduction with psychomotor impairment in methamphetamine abusers. *Am. J. Psychiatry*. 158:377, 2001.

235. Worsley, J.N., Moszczynska, A., Falardeau, P. et al. Dopamine D1 receptor protein is elevated in nucleus accumbens of human, chronic methamphetamine users. *Mol. Psychiatry*. 5:664, 2000.

236. Tong, J., Ross, B.M., Schmunk, G.A. et al. Decreased striatal dopamine D1 receptor-stimulated adenylyl cyclase activity in human methamphetamine users. *Am. J. Psychiatry*. 160:896, 2003.

237. Frost, D.O., Cadet, J.-L. Effects of methamphetamine-induced neurotoxicity on the development of neural circuitry: a hypothesis. *Brain Res. Rev*. 34:103, 2000.

238. Cho, A.K., Melega, W.P. Patterns of methamphetamine abuse and their consequences. *J. Addict. Dis*. 21:21, 2002.

239. Cadet, J.-L., Jayanthi, S., Deng, X. Speed kills: cellular and molecular bases of methamphetamine-induced nerve terminal degeneration and neuronal apoptosis. *FASEB J*. 17:1775, 2003.

240. Kita, T., Wagner, G.C., Nakashima, T. Current research on methamphetamine-induced neurotoxicity: animal models of monoamine disruption. *J. Pharmacol. Sci*. 92:178, 2003.

241. Itzhak, Y., Achat-Mendes, C. Methamphetamine and MDMA (Ecstasy) neurotoxicity: "of mice and men." *IUBMB Life*. 56:249, 2004.

242. Meredith, C.W., Jaffe, C., Ang-Lee, K., Saxon, A.J. Implications of chronic methamphetamine use: a literature review. *Harvard Rev. Psychiatry*. 13:141, 2005.

243. Wagner, G.C., Ricaurte, G.A., Seiden, L.S. et al. Long-lasting depletions of striatal dopamine and loss of dopamine uptake sites following repeated administration of methamphetamine. *Brain Res*. 181:151, 1980.

244. Ricaurte, G.A., Guillery, R.W., Seiden, L.S., Schuster, C.R., Moore, R.Y. Dopamine nerve terminal degeneration produced by high doses of methylamphetamine in the rat brain. *Brain Res*. 235:93, 1982.

245. Ricaurte, G.A., Guillery, R.W., Seiden, L.S., Schuster, C.R. Nerve terminal degeneration after a single injection of D-amphetamine in iprindole-treated rats: relation to selective long-lasting dopamine depletion. *Brain Res*. 291:378, 1984.

246. Eisch, A.J., Marshall, J.F. Methamphetamine neurotoxicity: dissociation of striatal dopamine terminal damage from parietal cortical cell body injury. *Synapse.* 30:433, 1998.
247. Brunswick, D.J., Benmansour, S., Tejani-Butt, S.M., Hauptmann, M. Effects of high-dose methamphetamine on monoamine uptake sites in rat brain measured by quantitative autoradiography. *Synapse.* 11:287, 1992.
248. Villemagne, V., Yuan, J., Wong, D.F. et al. Brain dopamine neurotoxicity in baboons treated with doses of methamphetamine comparable to those recreationally abused by humans: evidence from [11C]WIN-35,428 positron emission tomography studies and direct *in vitro* determinations. *J. Neurosci.* 18:419, 1998.
249. Koda, L.Y., Gibb, J.W. Adrenal and striatal tyrosine hydroxylase activity after methamphetamine. *J. Pharmacol. Exp. Ther.* 185:42, 1973.
250. Trulson, M.E., Cannon, M.S., Faegg, T.S., Raese, J.D. Effects of chronic methamphetamine on the nigral-striatal dopamine system in rat brain: tyrosine hydroxylase immunochemistry and quantitative light microscopic studies. *Brain Res. Bull.* 15:569, 1985.
251. Ricaurte, G.A., Schuster, C.R., Seiden, L.S. Long-term effects of repeated methylamphetamine administration on dopamine and serotonin neurons in the rat brain: a regional study. *Brain Res.* 193:153, 1980.
252. Frey, K., Kilbourn, M., Robinson, T. Reduced striatal vesicular monoamine transporters after neurotoxic but not after behaviorally-sensitizing doses of methamphetamine. *Eur. J. Pharmacol.* 334:273, 1997.
253. Bowyer, J.F., Tank, A.W., Newport, G.D. et al. The influence of environmental temperature on the transient effects of methamphetamine on dopamine levels and dopamine release in rat striatum. *J. Pharmacol. Exp. Ther.* 260:817, 1992.
254. Ali, S.F., Newport, G.D., Holson, R.R., Slikker, W.J., Bowyer, J.F. Low environmental temperatures or pharmacologic agents that produce hypothermia decrease methamphetamine neurotoxicity in mice. *Brain Res.* 58:33, 1994.
255. Harvey, D.C., Lacan, G., Tanious, S.P., Melega, W.P. Recovery from methamphetamine induced long-term nigrostriatal dopaminergic deficits without substantia nigra cell loss. *Brain Res.* 871:259, 2000.
256. Harvey, D.C., Lacan, G., Melegan, W.P. Regional heterogeneity of dopaminergic deficits in vervet monkey striatum and substantia nigra after methamphetamine exposure. *Exp. Brain Res.* 133:349, 2000.
257. Seiden, L.S., Fischman, M.W., Schuster, C.R. Long-term methamphetamine induced changes in brain catecholamines in tolerant rhesus monkeys. *Drug Alcohol Depend.* 1:215, 1976.
258. Melega, W.P., Lacan, G., Harvey, D.C., Huang, S.C., Phelps, M.E. Dizocilpine and reduced body temperature do not prevent methamphetamine-induced neurotoxicity in the vervet monkey: [11C]WIN 35,428-positron emission tomography studies. *Neurosci. Lett.* 258:17, 1998.
259. Woolverton, W.L., Ricaurte, G.A., Forno, L.S., Seiden, L.S. Long-term effects of chronic methamphetamine administration in rhesus monkeys. *Brain Res.* 486:73, 1989.
260. Melega, W.P., Raleigh, M.J., Stout, D.B. et al. Recovery of striatal dopamine function after acute amphetamine- and methamphetamine-induced neurotoxicity in the vervet monkey. *Brain Res.* 766:113, 1997.
261. Cass, W.A., Manning, M.W. Recovery of presynaptic dopaminergic functioning in rats treated with neurotoxic doses of methamphetamine. *J. Neurosci.* 19:7653, 1999.
262. Cass, W.A. Attenuation and recovery of evoked overflow of striatal serotonin in rats treated with neurotoxic doses of methamphetamine. *J. Neurochem.* 74:1079, 2000.
263. Friedman, S.D., Castaneda, E., Hodge, G.K. Long-term monoamine depletion, differential recovery, and subtle behavioral impairment following methamphetamine-induced neurotoxicity. *Pharmacol. Biochem. Behav.* 61:35, 1998.
264. Melega, W.P., Quintana, J., Raleigh, M.J. et al. 6-[18F]Fluoro-L-DOPA-PET studies show partial reversibility of long-term effects of chronic amphetamine in monkeys. *Synapse.* 22:63, 1996.
265. Moszczynska, A., Fitzmaurice, P., Ang, L. et al. Why is parkinsonism not a feature of human methamphetamine users? *Brain.* 127:363, 2003.
266. Wilson, J.M., Kalasinsky, K.S., Levey, A.I. et al. Striatal dopamine nerve terminal markers in human, chronic methamphetamine users. *Nat. Med.* 2:699, 1996.

267. Volkow, N.D., Chang, L., Wang, G.J. et al. Higher cortical and lower subcortical metabolism in detoxified methamphetamine abusers. *Am. J. Psychiatry.* 158:383, 2001.

268. Sonsalla, P.K., Nicklas, W.J., Heikkila, R.E. Role for excitatory amino acids in methamphetamine-induced nigrostriatal dopaminergic toxicity. *Science.* 243:398, 1989.

269. Gibb, J.W., Kogan, F.J. Influence of dopamine synthesis on methamphetamine-induced changes in striatal and adrenal tyrosine hydroxylase activity. *Naunyn Schmiedeberg's Arch. Pharmacol.* 310:185, 1979.

270. Schmidt, C.J., Ritter, J.K., Sonsalla, P.K., Hanson, G.R., Gibb, J.W. Role of dopamine in the neurotoxic effects of methamphetamine. *J. Pharmacol. Exp. Ther.* 233:539, 1985.

271. Johnson, M., Stone, D.M., Hanson, G.R., Gibb, J.W. Role of the dopaminergic nigrostriatal pathway in methamphetamine-induced depression of the neostriatal serotonergic system. *Eur. J. Pharmacol.* 135:231, 1987.

272. Seiden, L.S., Vosmer, G. Formation of 6-hydroxydopamine in caudate nucleus of the rat brain after a single large dose of methylamphetamine. *Pharmacol. Biochem. Behav.* 21:29, 1984.

273. Axt, K.J., Commins, D.L., Vosmer, G., Seiden, L.S. alpha-Methyl-*p*-tyrosine pretreatment partially prevents methamphetamine-induced endogenous neurotoxin formation. *Brain Res.* 515:269, 1990.

274. DeVito, M.J., Wagner, G.C. Methamphetamine-induced neuronal damage: a possible role for free radicals. *Neuropharmacology.* 28:1145, 1989.

275. Cadet, J.L., Sheng, P., Ali, S. et al. Attenuation of methamphetamine-induced neurotoxicity in copper/zinc superoxide dismutase transgenic mice. *J. Neurochem.* 62:380, 1994.

276. Hirata, H., Ladenheim, B., Carlson, E., Epstein, C., Cadet, J.L. Autoradiographic evidence for methamphetamine-induced striatal dopaminergic loss in mouse brain: attenuation in CuZn-superoxide dismutase transgenic mice. *Brain Res.* 714:95, 1996.

277. Maragos, W.F., Jakel, R., Chesnut, D. et al. Methamphetamine toxicity is attenuated in mice that overexpress human manganese superoxide dismutase. *Brain Res.* 878:218, 2000.

278. Itzhak, Y., Gandia, C., Huang, P.L., Ali, S.F. Resistance of neuronal nitric oxide synthase-deficient mice to methamphetamine-induced dopaminergic neurotoxicity. *J. Pharmacol. Exp. Ther.* 284:1040, 1998.

279. Fumagalli, F., Gainetdinov, R.R., Wang, Y.M. et al. Increased methamphetamine neurotoxicity in heterozygous vesicular monoamine transporter 2 knock-out mice. *J. Neurosci.* 19:2424, 1999.

280. Wagner, G.C., Carelli, R.M., Jarvis, M.F. Ascorbic acid reduces the dopamine depletion induced by methamphetamine and the 1-methyl-4-phenyl pyridinium ion. *Neuropharmacology.* 25:559, 1986.

281. Huang, N.K., Wan, F.J., Tseng, C.J., Tung, C.S. Nicotinamide attenuates methamphetamine-induced striatal dopamine depletion in rats. *Neuroreport.* 8:1883, 1997.

282. Ali, S.F., Martin, J.L., Black, M.D., Itzhak, Y. Neuroprotective role of melatonin in methamphetamine- and 1-methyl-4-phenyl-1,2,3,6-tetrahydropyridine-induced dopaminergic neurotoxicity. *Ann. N.Y. Acad. Sci.* 890:119, 1999.

283. Hirata, H., Asanuma, M., Cadet, J.L. Melatonin attenuates methamphetamine-induced toxic effects on dopamine and serotonin terminals in mouse brain. *Synapse.* 30:150, 1998.

284. Imam, S.Z., Newport, G.D., Islam, F., Slikker, W.J., Ali, S.F. Selenium, an antioxidant, protects against methamphetamine-induced dopaminergic neurotoxicity. *Brain Res.* 18:575, 1999.

285. Imam, S.Z., Ali, S.F. Selenium, an antioxidant, attenuates methamphetamine-induced dopaminergic toxicity and peroxynitrite generation. *Brain Res.* 855:186, 2000.

286. Kim, H.C., Jhoo, W.K., Choi, D.Y. et al. Protection of methamphetamine nigrostriatal toxicity by dietary selenium. *Brain Res.* 851:76, 1999.

287. London, E.D., Simon, S.L., Berman, S.M. et al. Mood disturbances and regional cerebral metabolic abnormalities in recently abstinent methamphetamine abusers. *Arch. Gen. Psychiatry.* 61:73, 2004.

288. Iyo, M., Namba, H., Yanagisawa, M. et al. Abnormal cerebral perfusion in chronic methamphetamine abusers: a study using 99MTc-HMPAO and SPECT. *Prog. Neuropsychopharmacol. Biol. Psychiatry.* 21:789, 1997.

289. Alhassoon, O.M., Dupont, R.M., Schweinsburg, B.C., et al. Regional cerebral blood flow in cocaine- versus methamphetamine-dependent patients with a history of alcoholism. *Int. J. Neuropsychopharmacol.* 4:105, 2001.

290. Wang, G.J., Volkow, N.D., Chang, L. et al. Partial recovery of brain metabolism in methamphetamine abusers after protracted abstinence. *Am. J. Psychiatry.* 31:313, 2004.

291. McCann, U.D., Wong, D.F., Yokoi, F. et al. Reduced striatal dopamine transporter density in abstinent methamphetamine and methcathinone users: evidence from positron emission tomography studies with [11C]WIN-35,428. *J. Neurosci.* 18:8417, 1998.

292. Sekine, Y., Iyo, M., Ouchi, Y. et al. Methamphetamine-related psychiatric symptoms and reduced brain dopamine transporters studied with PET. *Am. J. Psychiatry.* 58:1206, 2001.

293. Iyo, M., Sekine, Y., Mori, N. Neuromechanism of developing methamphetamine psychosis: a neuroimaging study. *Ann. N.Y. Acad. Sci.* 1025:288, 2004.

294. Volkow, N.D., Wang, G.J., Fowler, J.S. et al. Decreases in dopamine receptors but not in dopamine transporters in alcoholics. *Alcohol Clin. Exp. Res.* 20:1594, 1996.

295. Volkow, N.D., Fowler, J.S., Wolf, A.P. et al. Effects of chronic cocaine abuse on postsynaptic dopamine receptors. *Am. J. Psychiatry.* 147:740, 1990.

296. Wang, G.J., Volkow, N.D., Fowler, J.S. et al. Dopamine D2 receptor availability in opiate-dependent subjects before and after naloxone-precipitated withdrawal. *Neuropsychopharmacology.* 16:174, 1997.

297. Rawson, R.A., Gonzales, R., Brethen, P. Treatment of methamphetamine use disorders: an update. *J. Subst. Abuse Treat.* 23:145, 2002.

298. Cretzmeyer, M., Sarranzin, M.V., Huber, D.L., Block, R.I., Hall, J.A. Treatment of methamphetamine abuse: research findings and clinical directions. *J. Subst. Abuse Treat.* 24:267, 2003.

CHAPTER 5

Neurochemical Adaptations and Cocaine Dependence

Kelly P. Cosgrove, Ph.D. and Julie K. Staley, Ph.D.
Department of Psychiatry, Yale University School of Medicine, New Haven, Connecticut and VA Connecticut Healthcare System, West Haven, Connecticut

CONTENTS

Cocaine addiction is a disease of the brain that is characterized by compulsive drug-taking behavior that may be a consequence of an altered neurochemical state. Cocaine use by inhalation or injection produces a rapid, intense, euphoric "rush," which leads to repeated use. The repeated use of cocaine causes long-lasting changes in brain receptors and transporters that perpetuate drug use and the resulting dysphoria or depression. Furthermore, regulatory changes in neurochemical targets that have occurred as a compensatory response to "oppose" or "neutralize" the pharmacological effects of the drug persist after the drug has cleared from the brain, and may underlie the craving and dysphoria associated with cocaine withdrawal and relapse. Revolutionary advances in basic neuroscience have catalyzed extraordinary efforts toward the discovery and development of novel central nervous system (CNS) drugs effective for the treatment of cocaine dependence. These advances have included the molecular cloning and characterization of many of the receptors and transporters implicated in the rewarding effects of and dependence on cocaine. Understanding how and which receptors and transporters are altered by chronic cocaine use may identify targets for the development of drugs that will alleviate the symptoms associated with the initiation and perpetuation of drug-taking behavior. Current drug discovery efforts have taken multiple approaches toward the development of cocaine pharmacotherapies. Many of the novel pharmacological agents currently under development are directed toward a single molecular target related to or regulated by cocaine. Some pharmacotherapies currently under evaluation are directed toward multiple distinct receptor and/or transporter populations known to modulate the activity of the drug reward circuit. Several approaches have been taken to develop medications to treat cocaine dependence. Cocaine "substitute" medications are drugs that act via a similar mechanism as cocaine but with a more limited abuse potential. This strategy seeks to alleviate some of the withdrawal effects and reduce craving and is similar to nicotine replacement therapy for tobacco smoking or methadone maintenance for heroin addiction. Cocaine antagonists block the effects of cocaine or block the binding of cocaine to the dopamine transporter and thus attenuate the reinforcing effects of cocaine and reduce the likelihood of cocaine use. This is similar to naltrexone therapy for heroin addiction in which naltrexone, a mu-opioid receptor antagonist, blocks the binding of heroin or other opiates to the mu-opioid receptor. This strategy may also target other receptor systems that modulate dopaminergic activity to functionally antagonize the effects of cocaine. The cocaine vaccine interferes with the pharmacokinetics of cocaine by blocking or slowing cocaine's uptake into the brain, thus eliminating its reinforcing effects. Thus far, several promising treatments have been developed yet no truly effective medications exist and none is approved specifically for the treatment of cocaine dependence. This chapter reviews recent research studies that have identified neurochemical targets regulated by chronic cocaine use and their implications for the development of pharmacotherapies for cocaine dependence.

5.1 NEUROCHEMISTRY OF COCAINE DEPENDENCE

Many drugs with abuse liability including cocaine have been shown to enhance dopaminergic neurotransmission in the mesolimbic drug reward circuits. Cocaine, an indirect-acting dopaminergic agonist, binds to recognition sites on the plasma membrane dopamine (DA) transporter and increases dopamine levels by preventing the reuptake of released dopamine.[1–3] Intravenous injection of cocaine has been shown to significantly inhibit dopamine reuptake within 4 s.[4] The reinforcing effects of cocaine are initiated by the interactions of dopamine with pre- and postsynaptic DA receptors. Two classes of DA receptors have been classified including the D_1 receptor family (D_1, D_{1a}, and D_5/D_{1b} receptor subtypes) and the D_2 receptor family (D_2, D_3, and D_4 receptor subtypes).[5] The DA receptor subtypes are distinguished by their distinct anatomical, molecular, pharmacological, and signal transduction properties.[5] When cocaine is present, extracellular dopamine levels are elevated resulting in chronic stimulation of the DA receptors.[3] This persistent interaction of dopamine with its receptors alters the DA receptor signaling, which may, in part, underlie the reinforcing properties of cocaine.[3] Furthermore, postsynaptic DA receptors have been localized to other neurotransmitter-containing pathways including GABAergic, glutamatergic, and cholinergic projections indicating that additional neurochemical pathways may undergo compensatory adaptive changes in response to the persistent activation of DA receptors. The effects of cocaine in brain are widespread in the mesolimbic system including the ventral tegmental area, ventral striatum, and prefrontal cortex.

Recently, considerable evidence has accumulated suggesting that other neurochemical substrates (i.e., glutamatergic, serotonergic, GABAergic, and opioidergic) also play a role in the development of cocaine dependence. The anatomical organization of these neurochemical systems within the drug reward circuit suggests that functional interactions may occur between dopaminergic systems and neighboring neural pathways. Nigrostriatal dopaminergic neurons and corticostriatal glutamatergic neurons colocalize on common GABAergic medium spiny dendrites in the striatum suggesting the potential for interdependency between the two circuits.[6] Glutamate antagonists that act at the NMDA receptor complex block stereotypy and locomotor activation in animal models of cocaine dependence indicating that the glutamatergic pathway may be critical to the expression of these psychomotor stimulant behaviors.[6,7] Serotonergic projections to the ventral tegmental area input onto dopamine cell bodies suggesting that serotonin modulates mesolimbic DA neurotransmission by direct stimulation (i.e., cocaine increases extracellular serotonin by blocking reuptake) or by indirect modulatory feedback mechanisms from DA nerve terminal activation.[8] Mesolimbic DA neurotransmission is modulated in part by tonic activation of the μ- and κ-opioid receptors located in the vicinity of the DA cell bodies and DA nerve terminals, respectively. Activation of μ-opioid receptors in the ventral tegmental area increases dopamine release in the nucleus accumbens, whereas activation of the κ-opioid receptors in the nucleus accumbens inhibits dopamine release.[9–11] Cross-talk between opioids, serotonin, glutamate, GABA, and dopamine may lead to a sequence of neuroadaptive processes that contribute to the behavioral and physiological manifestations associated with cocaine dependence. Additionally, understanding of the neuroadaptive changes involved in cocaine dependence may lead to the development of effective medications for the treatment of the disorder.

5.2 DA TRANSPORTER AND COCAINE DEPENDENCE

5.2.1 Regulation of the Dopamine Transporter (DAT) by Cocaine

The addictive liability of cocaine and other DA-enhancing psychostimulants may be related to compensatory adaptation of the dopamine transporter (DAT) to chronic elevations of intrasynaptic DA. The DAT is a protein on neuronal membranes that is involved in synaptic transmission by transporting DA from the extracellular space to the neuron.[12] There is a wealth of evidence suggesting that cocaine mediates its powerful reinforcement by binding to recognition sites on the DAT. Interestingly, mice lacking the DAT (DAT-KO) still self-administer cocaine[13] and show cocaine-induced conditioned place preference,[14] indicating that cocaine exerts strong reinforcing effects via alternative neurotransmitter systems. Persistent inhibition of DA reuptake by cocaine has been shown to alter the number of cocaine recognition sites associated with the DAT. Chronic treatment of rats with intermittent doses of cocaine resulted in a two- to fivefold increase in the apparent density of [³H]cocaine and [³H]BTCP binding in the striatum.[15] Significant increases in striatal [³H]WIN 35,428 binding were observed in rats allowed to self-administer cocaine in a chronic unlimited access paradigm,[16] rats treated intermittently and continuously with cocaine,[17] and rabbits treated with intermittent cocaine injections.[18] Chronic cocaine exposure in nonhuman primates resulted in significantly increased DAT binding sites compared to binding in drug-naïve control monkeys in the ventral striatum measured with [³H]WIN 35,248.[19]

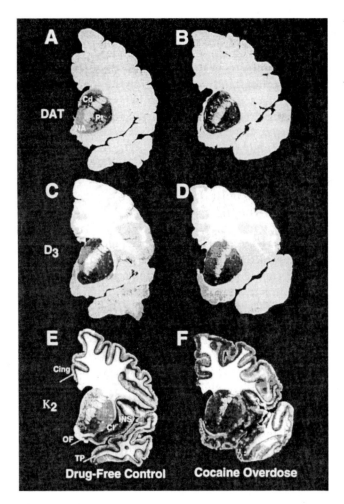

Figure 5.1   Visualization of the distribution of the DA transporter, D_3 receptor, and κ_2-opioid receptor in the human brain of a drug-free control subject and a representative cocaine overdose victim. (A, B) The DA transporter was measured using [^{3}H]WIN 35,428 (2 nM) as described previously. (C, D) The D_3 receptor was measured using [3H]-(+)-7-OH-DPAT (1 nM) in the presence of GTP (300 mM) to enhance the selective labeling of the D_3 receptor subtype over the D_2 receptor subtype as described previously. (E, F) The κ_2-opioid receptor subtype was measured using [^{125}I]IOXY on tissue sections pretreated with BIT and FIT to occlude binding to the μ- and δ-opioid receptors, respectively.

Consistently, in post-mortem brain from human cocaine users, striatal DAT binding sites were significantly increased compared to controls using [^{3}H]WIN 35,248.[20,21] DAT densities detected using [^{125}I]RTI-55 and [^{3}H]WIN 35,248 were elevated throughout the caudate, putamen, and nucleus accumbens of cocaine-related deaths[22] and fatal cocaine overdose victims (Figure 5.1).[22–25] Furthermore, Rosenthal plots of the saturation binding data demonstrated that the increase of [^{3}H]WIN 35,428 binding observed in the cocaine overdose victims was due to an elevation in the apparent density of the high-affinity cocaine recognition site on the DAT.[24] While most studies indicate chronic cocaine results in increased DAT binding sites, some have reported a decrease[26] or no change.[27] Further, it was reported that administration of GBR 12909 (also called vanoxerine) but *not* cocaine resulted in a decrease in DAT density in caudate putamen and nucleus accumbens[28] highlighting that while GBR 12909 is a selective DA reuptake blocker, cocaine's effect involves other transporter systems. Importantly, elevations in DAT demonstrated in the post-mortem brains of human cocaine abusers are supported by *in vivo* SPECT imaging in human cocaine-dependent subjects.[29] Here, the striatal uptake of [^{123}I]β-CIT (also called RTI-55) was significantly elevated (25%) in acutely abstinent (≥96 h) cocaine-dependent subjects. These studies suggest that the high-affinity cocaine binding sites are upregulated in the striatal reward centers with chronic cocaine use as a compensatory response to elevated synaptic levels of DA. It was hypothesized that this upregulation of cocaine recognition sites associated with the DAT may reflect an increased ability of the protein to transport DA. As the transporter elevates its apparent density in the nerve terminal to clear DA from the synapse, more cocaine will be needed to experience cocaine's reinforcing effects and euphoria.[24] This neuroadaptive regulation of the DAT may occur as a result of the direct interaction of cocaine with the DAT, or, alternatively, may be due to feedback mechanisms that are activated as a consequence of prolonged elevation concentrations of DA. Recent evidence suggests that chronic cocaine may also lead to increased DAT function. Mash and colleagues[30] reported increased [^{3}H]WIN 34,428 binding and [^{3}H]dopamine uptake in post-mortem human striatal synaptosomes from cocaine addicts compared to age-matched controls, indicating that chronic cocaine functionally upregulates DAT. Additionally, evidence from cell cultures suggests that the increase in DAT transport activity from chronic cocaine may be due to the concurrent increase in DAT cell surface expression.[31,32] The direct regulation of DAT by cocaine suggests that the DAT is an ideal target for the development of anti-cocaine medications.

5.2.2 Rate of Interaction of Cocaine with the DAT and Cocaine Reinforcement

An alternative neurochemical hypothesis for the reinforcing efficacy of cocaine at the DAT is related to the rate of entry of cocaine into the brain, coupled with its ability to rapidly bind to the DAT, inhibit DA uptake, and enhance dopaminergic neurotransmission.[3] This hypothesis is based on the knowledge that other drugs known to block DA uptake, such as mazindol, GBR 12909, and methylphenidate, are not as reinforcing as cocaine[33] and exhibit a slower rate of entry into the brain and a slower onset of action.[34] Furthermore, GBR 12909, methylphenidate, bupropion, and mazindol displace the *in vivo* binding of the radiolabeled cocaine congener [^{3}H]WIN 35,428 at a considerably slower rate than cocaine.[35] In humans, methylphenidate demonstrates a lower abuse liability as compared to cocaine, but enters the CNS with a rate similar to cocaine. An *in vivo* study determined that the potency of methylphenidate vs. cocaine at the DAT is similar and thus unlikely to underlie the differences in abuse potential.[36] However, methylphenidate is cleared more slowly than cocaine and has a longer duration of side effects, and both factors may underlie its decreased rate of administration and abuse liability.[37] Imaging studies have shown that, in general, while the level of DAT blockade is important in determining the intensity of the euphoric effects or "high," the rate at which substances block DAT determines the perceived "high."[38] For example, in monkeys, a PET study showed that cocaine and GBR 12909 affected DA synthesis and DAT availability with different time courses; e.g., GBR 12909 decreased [^{11}C]β-CFT binding for a longer time than did cocaine.[39] These studies suggest that pharmacological interventions that decrease the rate of entry into or clearance of cocaine from the brain, or alternatively block or slow cocaine's interaction with the DA transporter may be useful for the treatment of cocaine dependence.

5.2.3 The DAT as a Target for Cocaine Pharmacotherapies

One pharmacotherapy strategy for the treatment of cocaine dependence that has received significant attention is the development of drugs that antagonize or substitute for cocaine at its site of action in the brain.[40,41] Hypothetically, the "ideal cocaine antagonist" would manifest high-affinity binding to cocaine recognition sites on the DAT, slow dissociation from these binding sites, minimal inhibition of substrate binding and uptake, a long biological half-life, and low abuse liability.[40–42] The DA reuptake inhibitor, GBR 12909, appears to satisfy several of these criteria.[40–42]

At low doses, GBR 12909 binds to the DAT with high affinity, dissociates slowly, causes only a modest increase in extracellular DA, and partially antagonizes the increase in extracellular DA evoked by local perfusion of cocaine into the striatum and the nucleus accumbens.[40,41] However, at high doses, GBR 12909 produces locomotor activation, stereotypy, and behavioral sensitization, and cross-sensitizes with cocaine.[42] Importantly, behavioral studies indicate that GBR 12909 reduces cocaine-maintained behaviors in rats[43] and in nonhuman primates.[44] The dose of GBR 12909 that reduced cocaine self-administration in rhesus monkeys[44] was quantified *in vivo* in baboon and results indicated that GBR 12909 must occupy at least 50% of DAT to translate into behavioral efficacy, e.g., reduce responding for cocaine.[45] This is consistent with the finding that doses of cocaine that are reliably self-administered by rhesus monkeys occupy 53 to 87% of DAT.[46] These studies suggest that the development of a cocaine antagonist is plausible; however, efficacy as a cocaine antagonist may be dose-related. At some doses GBR 12909 has also been shown to *increase* or reinstate cocaine-seeking behavior,[43] but it is unclear whether this would translate into high abuse liability in a human clinical population. Other DAT inhibitors that have been shown to reduce the self-administration of cocaine in animals include PTT, RTI 113, and HD-23.[47–49] Molecular characterization studies of DAT, which used chimeric dopamine-norepinephrine transporters, delineated discrete domains for substrate and cocaine interactions.[50,51] These studies support the development of a cocaine antagonist devoid of uptake blockade activity for the clinical management of cocaine addiction. Current drug discovery efforts have focused on the development of compounds using cocaine as the core structure, in an effort to find a drug that blocks cocaine interactions with the DAT, but does not block the normal uptake function of the transporter. While these studies are in the early stages, it is anticipated that this approach may lead to the development of an efficacious anti-cocaine medication.

5.2.4   Cocaine Vaccines

An alternative pharmacotherapy for cocaine dependence currently under investigation is use of a cocaine vaccine to blunt the reinforcing effects of cocaine.[51–60] The basis of this pharmacotherapy is to decrease the rate of entry of cocaine into the CNS (and therefore the onset of action), by either binding cocaine with antibody generated by active immunization with a stable cocaine conjugate or by using an enzymatically active antibody specific for cocaine.

Because cocaine is a small molecule (MW = 303 g/mol) it is unlikely that it will be immunogenic and therefore must be conjugated to a carrier molecule such as KLH (keyhole limpet hymacyanin), polyethylene glycol, diphtheria, or tetanus toxoids to enhance its immunogenicity.[53–55] While early attempts to make a cocaine vaccine did not demonstrate significant efficacy for blocking the effects of cocaine in the CNS,[55] recent studies have been more successful.[56,57] Active immunization with a stable cocaine-KLH conjugate lowered cocaine levels in the brain by 80% in immunized vs. control rats, enough to decrease cocaine-induced hyperlocomotion and stereotypic behavior.[56] However, while the active immunization approach to cocaine pharmacotherapy appears promising, its success will be hindered if the addict administers large enough quantities of cocaine to override the antibody-induced blockade.[57] Active immunization with the cocaine immunogen GNC-KLH significantly reduced cocaine-induced reinstatement in rats and resulted in an eightfold rightward shift in the cocaine dose–effect curve, indicating that surmountability of the vaccine occurred at a dose of cocaine that was eight times higher than a dose necessary to maintain cocaine responding.[61] Subsequently, a second-generation hapten 3 (GND) was synthesized and, in rats, active immunization with GND-KLH resulted in robust decreases in cocaine-induced ambulatory and stereotypic behaviors.[62]

Use of a catalytic cocaine antibody may bypass this potential downfall of the active immunization approach. The enzymatically active catalytic cocaine antibody cleaves cocaine into two inactive metabolites: ecognine methyl ester and benzoic acid. The two metabolites are released from the catalytic antibody rendering the antibody free to degrade more cocaine. Because the catalytic antibodies are not depleted, it is impossible for the cocaine abuser to "override" the presence of the antibody by administering higher doses of cocaine.[58] Catalytic antibodies have been generated against transition state analogues of cocaine[58–60] and a catalytic antibody was shown to reduce cocaine-induced toxicity and block the self-administration of cocaine in rats.[63] While this strategy to blunt the reinforcing effects of cocaine is promising, the antibody will not alleviate the craving, dysphoria, anxiety, or depression often linked to relapse. Thus, although the cocaine antibodies may prevent cocaine reinforcement, they will not block the reinforcing effects of other psychostimulants that an addict may administer to relieve withdrawal symptoms.[64] In addition, the success of the catalytic antibody may be hindered if the antibody is perceived as foreign, and idiotypic antibodies are produced that will interact with the cocaine antibody, rendering it enzymatically inactive.

The safety of the cocaine vaccine TC-CD in former cocaine abusers has been evaluated in a Phase I clinical trial, and it was determined that the vaccine was well tolerated with dose-related increases in antibody levels.[65] Two Phase II clinical trials have now been conducted.[66,67] The vaccine was again well tolerated and subjects reported a reduction in cocaine's reinforcing effects. The antibody levels were detectable after the second dose, peaked at 8 to 12 weeks, and remained elevated for up to 6 months; preliminary findings indicated a negative association between antibody level and cocaine use. Other anti-cocaine vaccines in development include a blocking antibody (ITAC-cocaine) and a monoclonal catalytic antibody (15A10).

5.3 DA RECEPTORS AND COCAINE DEPENDENCE

The rewarding effects of dopamine are mediated by five DA receptor subtypes distinguished by their unique molecular and pharmacological properties and distinct anatomical locations. Repeated and prolonged elevations in synaptic dopamine levels that result from the binge use of cocaine may result in alterations in the affinity, number, or coupling state of the DA receptors. At present, the relative contribution of each of the DA receptor subtypes to the rewarding effects of cocaine is not clear. Dopamine agonists that interact with receptors belonging to both the D_1 and D_2 receptor families function as positive reinforcers, while both D_1-like and D_2-like receptor antagonists decrease the reward value of psychostimulants.[68,69] Stimulation of D_1 or D_2 receptors in the ventral tegmental area enhances the rewarding effect for brain stimulation.[70] These findings suggest that compensatory changes in DA receptor number or signaling may contribute to the development of cocaine dependence.

5.3.1 Cocaine-Induced D_1 Receptor Adaptations

The reinforcing effects of cocaine are mediated, in part, by the D_1 receptors in the nucleus accumbens and the central nucleus of the amygdala.[71] In preclinical animal studies, administration of the D_1 receptor antagonist SCH 23390 prevents reinstatement to cocaine-induced conditioned place preference,[72] reduces reinstatement to cocaine-seeking behavior,[73] and SCH 23390 when administered in combination with cocaine prevents the development of cocaine sensitization.[74,75] While cocaine self-administration is increased in the presence of D_1 receptor antagonists,[68,76–79] which is likely a compensatory response, it has recently been demonstrated that, under some schedules of reinforcement, low doses of benzazepine D_1 receptor antagonists block cocaine self-administration.[80] Additionally, a variety of D_1 receptor agonists (SKF 82958, SKF 81297, SKF 83959) and an antagonist (ecopipam) reduced reinstatement of cocaine-seeking in nonhuman primates.[81] Further, SCH 23390 reversed the attenuating effects of SKF 81297 on cocaine-seeking behavior but not cue-induced reinstatement in rats[82] indicating that the reductions in cocaine-seeking behavior are not due simply to behavioral disruption. These findings suggest that the state of the D_1 receptor population may dictate the ability of D_1 receptor antagonists to enhance or to block the reinforcing effects of cocaine. D_1 receptor mRNA levels, receptor number, and receptor sensitivity are altered as a consequence of protracted cocaine exposure and may, in part, account for some of the D_1 receptor-mediated behavioral responses to cocaine. D_1 receptor mRNA is not altered in the striatum, nucleus accumbens, or substantia nigra of human cocaine abusers;[83] however, elevations in D_1 receptor mRNA were observed in the striatum of rats chronically treated with cocaine.[84] The D_1 receptor is also critical for mediating cocaine-induced expression of certain genes (the fos and Jun family immediate early genes which encode transcription factors) and molecules such as brain-derived neurotrophic factor, β-catenin, and Gαolf, which are implicated in mediating cocaine's actions in the nucleus accumbens and caudate putamen.[85] Increased D_1 receptor density was reported in the nucleus accumbens and olfactory tubercle in rat brain following twice-daily injections of cocaine.[86] Chronic cocaine treatment results in an adaptive increase in D_1 receptor number in olfactory tubercle, nucleus accumbens, central pallidum, and substantia nigra, which normalized within 1 day[87] and remained at baseline values for up to 3 days[88] and 30 days[89] after cocaine treatment. Conversely, decreased striatal D_1 receptor densities that persisted for at least 2 weeks after the last administration of cocaine have been observed.[84] Chronic cocaine exposure in monkeys resulted in increased D_1 receptor density in striatum.[90] Electrophysiological sensitivity of D_1 receptors in the nucleus accumbens neurons is enhanced in rats 2 weeks post-cocaine administration,[91,92] suggesting that despite the number of D_1 receptors detected, adaptation in the D_1 receptor signaling cascade may occur to enhance D_1-receptor-mediated dopamine transmission. When these findings are viewed collectively, it is difficult to ascertain precisely the regulatory adaptations that the D_1 receptor undergoes in response to protracted cocaine use. The differences may be due to a variety of factors including variations in dose, frequency, and route of administration. These factors have been shown to influence the adaptive responses of dopaminergic synaptic markers; therefore, the differences may be attributed to distinct drug administration protocols.[17] Regardless of the adaptive response observed, it is evident from these studies that the adaptations of the D_1 receptor may be integral to the development of cocaine dependence and therefore may be a useful target for the development of cocaine pharmacotherapies.

5.3.2   Cocaine-Induced Adaptations in D_2 Receptors

The actions of cocaine on D_2 receptors have been shown to be essential to the development of cocaine dependence. The D_2 receptor antagonist haloperidol inhibits the development of behavioral sensitization to cocaine.[74] Sensitization is believed to play an important role in drug craving and the reinstatement of compulsive drug-taking behavior;[93,94] thus, altered regulation of D_2 receptors may contribute to the reinforcing potential of cocaine. D_2 mRNA levels were not altered in the striatum of human cocaine abusers[83] or rats treated with cocaine.[95,96] D_2 receptor mRNA levels were decreased in the olfactory tubercle of rats treated with a single injection of cocaine,[95] whereas D_2 receptor mRNA levels were transiently elevated in rats treated chronically with cocaine.[84] A transient increase in the binding of [^{3}H]raclopride in the olfactory tubercle and rostral nucleus accumbens and caudate-putamen was observed after binge administration of cocaine[97] and elevations in [^{3}H]spiperone and [^{125}I]spiperone binding were seen in the nucleus accumbens, olfactory tubercle, and substantia nigra after intermittent cocaine administration.[87] In contrast, D_2 receptor densities were not significantly affected in the rat striatum 3 days post-cocaine administration.[88] In monkeys, chronic cocaine exposure resulted in a significant decrease in D_2 receptor density throughout the striatum measured with [^{3}H]raclopride.[90] Furthermore, [^{18}F]N-methylspiroperidol labeling in living cocaine abusers demonstrated decreased D_2 receptor densities after a 1-week detoxification[98] and quantitative immunoblotting of post-mortem brain samples demonstrated a decrease in protein levels of D_2 receptors in the nucleus accumbens of cocaine users compared to controls.[99] Viewed collectively, these studies suggest that the D_2 receptor may undergo a transient elevation in the response to acute cocaine administration, which normalizes upon cocaine abstinence. However, caution should be taken in the interpretation of these data because these radioligands do not discriminate between D_2 and D_3 receptors and, therefore, identity of the regulated D_2 receptor subtype is not known.

5.3.3 Cocaine-Induced Adaptations in D_3 Receptors

Cocaine-induced adaptations at the D_3 receptor may mediate some of the reinforcing effects of cocaine. D_3 receptor-preferring agonists, although not self-administered by drug-naïve monkeys, are self-administered by monkeys previously trained to self-administer cocaine.[100] D_3 receptor-preferring agonists substitute for the discriminative stimulus effects of cocaine and produce place preference in rats and monkeys indicating that the D_3 receptor may mediate some of the subjective effects of cocaine.[101–103] These studies suggest that adaptations in the affinity, density, or molecular expression of the D_3 receptor induced by chronic cocaine use may underlie, in part, the development of cocaine dependence.

The D_3 receptor-preferring agonist [^{3}H]-(+)-7-OH-DPAT has been used in quantitative *in vitro* autoradiography studies to assess the status of D_3 receptors in human cocaine overdose (CO) victims[104] (Figure 5.1). Binding of [^{3}H]-(+)-7-OH-DPAT was elevated one- to threefold in the nucleus accumbens and ventromedial sectors of the caudate and putamen in CO victims as compared to drug-free and age-matched control subjects. D_3 receptor/cyclophilin mRNA ratios were also increased sixfold in the nucleus accumbens in CO vs. control subjects, indicating that cocaine exposure affects D_3 receptor mRNA expression. These findings were confirmed by saturation analysis of the [^{3}H]-(+)-7-OH-DPAT binding in membranes from the nucleus accumbens. The affinity for [^{3}H]-(+)-7-OH-DPAT binding was not different in the CO victims or the "Excited Delirium" (ED) victims as compared to drug-free control subjects. However, the saturation binding density for the CO victims (4.4 ± 0.4 pmol/g) when compared to the drug-free control subjects (3.1 ± 0.2 pmol/g) was significantly elevated.[105]

Interestingly, after 1-day withdrawal from chronic treatment with cocaine, increased D_3 receptor densities were observed in the striatum, while decreased D_3 receptor densities were observed in the nucleus accumbens of the rat brain.[106] In another study D_3 receptor binding was increased in the nucleus accumbens and ventral caudate putamen at 31 days after the last cocaine delivery but not at the 2-day or 8-day time points compared to controls.[107] Additionally, the density of D_3 receptors was increased in animals exposed to cocaine-associated stimuli.[108] It should be noted, however, that in the first study mentioned[106] the densities of the D_3 receptors were high in the dorsal striatum, as compared to the nucleus accumbens, which contrasts with the intense localization of the D_3 receptor mRNA in the nucleus accumbens of rat and human brain[109,110] and the higher D_3 receptor densities in the ventral striatum and the nucleus accumbens of the human brain.[105] In another study, D_3 receptor mRNA expression was not altered in human cocaine abusers.[111] These findings suggest that D_3 receptor mRNA and binding sites may be differentially regulated by cocaine exposure. A single cocaine exposure was recently shown to increase brain-derived neurotrophic factor (BDNF) mRNA in the prefrontal cortex of rats and was associated with a long-lasting increase in D_3 mRNA and D_3 protein in the nucleus accumbens.[112] Chronic treatment of C_6 glioma cells transfected with the D_3 receptor cDNA with DA agonists increased D_3 receptor densities, but did not change D_3 mRNA abundance.[113] The upregulation in D_3 receptor densities was blocked by treatment with cycloheximide, suggesting that the increase was mediated by increased protein synthesis. Changes in proteins, mRNA, and BDNF in addition to changes in D_3 density are likely associated with long-lasting cocaine-conditioned and cocaine-seeking behavior.

Alternatively, increased [^{3}H]-(+)-7-OH-DPAT binding may reflect a selective increase in one of the D_3 receptor isoforms. The D_3 receptor-specific probes may have hybridized to multiple alternative splice variants, including the truncated D_3 receptors.[114] Because DAergic ligands do not bind to the proteins generated from the truncated splice variants, regulation of message levels and binding site densities would be dissociated. The relative abundance of specific D_3 receptor isoforms may vary also with alterations in DA neurotransmission. While the biological significance and function of the D_3 receptor splice variants are not understood, it has been suggested that alternative splicing may regulate the relative abundance of the different D_3 receptor isoforms to differentially modulate D_3 receptor-mediated signaling.[115] Therefore, it may be suggested that the elevation in D_3 receptor density after chronic cocaine use may reflect a selective increase in one of the D_3 receptor isoforms. Additional studies are needed to determine if this regulatory pattern occurs in the human brain. D_3 receptor adaptations that result from repeated activation of the DA neurotransmission due to chronic "binge" use of cocaine may contribute to the development of cocaine dependence. This adaptive increase in the D_3 receptor may enhance the reinforcing effects of cocaine and contribute to the development of cocaine dependence. Recent development of potential *in vivo* agents for imaging the D_3 receptor in living humans will advance our understanding of the role of D_3 receptors in cocaine dependence.[116]

5.3.4   DA Receptors as Targets for Cocaine Pharmacotherapies

The search for pharmacotherapies for cocaine dependence has focused on drugs that target the DA receptors. Cocaine's reinforcing properties result from its ability to prolong the action of dopamine at the DA receptors in brain reward regions.[3] From this perspective it has been suggested that DA antagonists may block cocaine use by preventing the interaction of dopamine with its receptors, and therefore block reinforcement. In animal self-administration studies, both D_1 receptor (SCH 23390[68,76]) and D_2 receptor antagonists (pimozide,[117,118] sulpiride,[118] chlorpromazine,[119] spiperone,[76] metoclopramide,[118] pherphenazine[120]) increase cocaine self-administration by decreasing the reinforcing potential of cocaine. However, recent studies have shown that low doses of benzazepine D_1 receptor antagonists attenuate cocaine self-administration under certain schedules of reinforcement.[80] These findings suggest that certain doses of D_1 receptor antagonists may be efficacious for the treatment of cocaine dependence. Flupentixol, a dopamine antagonist with high affinity for the D_1 receptor, has demonstrated some efficacy for decreasing craving and increasing treatment retention in human cocaine abusers.[121] D_2 receptor antagonists (haloperidol and chlorpromazine) have been efficacious for the treatment of paranoia and psychosis but not craving in human cocaine abusers.[122,123] D_2 receptor antagonists also elicit significant adverse side effects such as abnormal movements and are not effective at reducing cocaine use even at high doses.[124] While DA antagonists may block the reinforcing efficacy of cocaine, use of DA antagonists as pharmacotherapies may be hampered by their propensity to enhance cocaine withdrawal symptoms.[64] Furthermore, compliance is hindered by the dysphoric and extrapyramidal side effects associated with the blockade of DA neurotransmission.[125]

DA receptor agonists also have been suggested as anti-cocaine medications because of their propensity to reduce craving that occurs during cocaine withdrawal. A recent study using pergolide, a D_2/D_3 receptor agonist, has demonstrated some efficacy for decreasing craving.[126] Bromocriptine, a D_2-like agonist, has undergone extensive evaluation as a treatment for cocaine dependence and appears to reduce craving in cocaine abusers.[127] However, its efficacy as a pharmacotherapy for cocaine dependence was weak.[128] Furthermore, studies using an indirect-acting DA agonist, amantadine, were not as successful as anticipated.[129] The poor outcome of these studies, which were conducted in the late 1980s, may be explained by a recent preclinical study which demonstrated that D_2-like agonists actually enhance cocaine-seeking behavior or "prime" the addict to initiate another binge use of cocaine[125,130] and D_2 receptor agonists are typically no longer developed as potential medications for cocaine dependence. However, cabergoline, a long-acting D_2 receptor agonist, has recently been evaluated in a Phase II trial for cocaine dependence and was more effective than placebo in reducing cocaine use as measured by negative urine screens for cocaine metabolites.[131] D_1-like agonists oppose cocaine-seeking behavior induced by cocaine itself,[130] suggesting that D_1-like receptor agonists may be efficacious for the treatment of craving in cocaine dependence. However, while D_1-like agonists may attenuate the craving associated with cocaine withdrawal and relapse and may not enhance cocaine-seeking behavior, they may be reinforcing and therefore at risk to be abused themselves.

The close association of the D_3 receptor with the striatal reward pathways and its selective distribution in the mesolimbic dopamine system suggest that drugs that target the D_3 receptor subtype may decrease the reinforcing effects of cocaine. Because D_3 receptors' densities elevate as cocaine dependence develops, this upregulation of D_3 receptors may contribute to the reinforcing effects of cocaine.[105] From this perspective, the development of drugs that block D_3 receptor function may be useful for the treatment of cocaine dependence. In keeping with this hypothesis, the D_3-selective antagonists (–)DS 121[125] and SB-277011A[132] attenuate cocaine self-administration in rats and block cocaine reinstatement.[133,134] Alternatively, agents that act as D_3 agonists or partial agonists may be used as substitutes to treat cocaine dependence.[135] The compound BP 897, a highly selective D_3 agonist, reduced cocaine seeking behavior in rats;[136] however, BP 897 also demonstrates antagonist properties at human D_3 and D_2 receptors,[137] suggesting that the cocaine-reducing effects may be due to antagonism at these sites. While these studies are encouraging, additional research is necessary to confirm the efficacy of either D_3 receptor antagonists or agonists as clinically useful pharmacotherapies. Several other agents that act within the dopamine system are being evaluated in clinical trials. These include disulfiram, which may act by increasing brain levels of dopamine and decreasing levels of norepinephrine, selegiline, which is an irreversible MAO inhibitor, and reserpine, a Rauwolfia alkaloid currently used as an antihypertensive agent, which acts to deplete dopamine, norepinephrine, and serotonin in presynaptic vesicles; see Gorelick et al.[138] for review.

5.4 KAPPA-OPIOID RECEPTORS

The endogenous opioidergic system has been implicated as a primary mediator of the behavioral and reinforcing effects of cocaine.[94] (See Reference 139 for more information on the interaction of cocaine with the opioid system including the mu- and delta-opioid receptors.) Pharmacological and molecular cloning studies have recently reported the existence of at least three subtypes of κ-opioid receptors.[140–147] Receptor mapping studies have demonstrated that both κ_1-opioid receptor and κ_2-opioid receptor subtypes are prevalent throughout the mesocorticolimbic pathways in the human brain.[26,148,149] One striking difference in the localization of the two subtypes is reflected by the intense localization of the κ_2-opioid receptor subtype in the ventral or "limbic" sectors of the striatum and the nucleus accumbens in the human brain.[149] Conversely, the κ_1-opioid receptor subtype may preferentially localize to the dorsal or "sensorimotor" areas of the human striatum.[26] Based on their neuroanatomical distribution in the human striatum, it may be hypothesized that the κ_1-opioid receptor subtype modulates motor functions, while the κ_2-opioid receptor subtype mediates emotional behaviors and affect. The anatomical localization of the κ-opioid receptor subtypes and their intimate association with dopaminergic reward pathways suggests that regulatory alterations in both κ_1- and κ_2-opioid receptors may be important in cocaine dependence.

5.4.1 Regulation of Kappa-Opioid Receptors by Cocaine

At present, the functional significance and relevance of each of the κ-opioid receptor subtypes in the CNS and their role in modulating the brain reward pathways with chronic substance abuse are not well understood. An adaptive increase in the density of κ-opioid receptors in guinea pig brain after chronic cocaine treatments was detected using the κ_1-selective radioligand [^{3}H]U69,593.[150] Furthermore, elevations in [^{125}I]Tyr^1-D-Pro^{10}-dynorphin A binding to κ-opioid receptors were observed within the dorsal and "motor" sectors of the striatum of human cocaine abusers.[26] Dynorphin A demonstrates higher affinity to the κ_1-opioid receptor subtype as compared to the κ_2-opioid receptor subtype; therefore, it may be suggested on the basis of occupancy that the elevated binding of [^{125}I]Tyr^1-D-Pro^{10}-dynorphin A observed in these studies may be due to recognition of the κ_1-opioid receptor subtype.[147,151,152] These findings combined with animal behavioral studies (e.g., cocaine place preference and self-administration) suggest a definitive role for the κ_1-opioid receptor in cocaine dependence. While these studies are reasonably conclusive, other studies are not, due to the use of radioligands, which lack selectivity between κ-opioid receptor subtypes. After chronic continuous exposure to cocaine,[153] elevated binding of the nonselective opioid agonist and antagonist ([³H]bremazocine and [³H]naloxone, respectively) was observed in the nucleus accumbens. Thrice daily injections of cocaine in rats resulted in increased κ-opioid receptor density in cingulate cortex, nucleus accumbens, and caudate putamen.[86] Furthermore, in rats treated with cocaine using a binge-administration paradigm, binding of [³H]bremazocine to κ-opioid receptors was increased.[154] Binding density of [³H]U-69593, a selective κ-opioid receptor ligand, was significantly higher in caudate putamen and nucleus accumbens after chronic cocaine infusion,[155] indicating increased numbers of κ-opioid receptors in brain areas associated with drug craving and reward.

It is interesting that in the same animal model, κ-opioid receptor mRNA was decreased in the substantia nigra, but not in the ventral tegmental area of cocaine-treated rats.[156] A recent study also reported κ-opioid receptor mRNA levels are decreased after cocaine self-administration in rats in the nucleus accumbens and ventral tegmental area.[157] The reasons for this disconnect between κ-opioid receptor binding and the κ-opioid receptor mRNA are not known. However, the discrepancy may be due to the detection of multiple κ-opioid receptor mRNAs or binding sites or to the binding of the [³H]bremazocine to a κ-opioid receptor subtype distinct from the κ-opioid receptor message that was measured. While the interpretation of these studies with regards to which κ-opioid receptor subtype was measured and the regulation of each subtype by cocaine is difficult at this time, recent advances in the cloning of these receptor subtypes and the development of subtype-specific radioligands will clarify these issues in the near future.

Recently, pharmacological binding assays to selectively label the κ_2-opioid receptor subtype have been developed using the opioid antagonist [¹²⁵I]IOXY in the presence of drugs occluding binding to the μ- and κ-opioid receptor subtypes.[151] This strategy was used in ligand binding and *in vitro* autoradiography assays to assess the regulation of the κ_2-opioid receptor subtype after cocaine exposure in post-mortem human brain (death was from cocaine overdose) (Figure 5.1).[149,158] Quantitative region-of-interest densitometric measurements of [¹²⁵I]IOXY binding demonstrated a twofold elevation in the anterior and ventral sectors of the caudate and putamen and in the nucleus accumbens of human cocaine overdose victims as compared to drug-free and age-matched control subjects. In subjects who experienced paranoia and agitation prior to their death, κ_2-opioid receptors were also elevated in the amygdala.[104,158]

The regulation of κ_2-opioid receptor numbers in the striatal reward centers suggests that adaptations in the κ_2-opioid receptors may also contribute to the development of cocaine dependence. κ-opioid agonists do not generalize to cocaine cues in drug discrimination paradigms;[159,160] however, κ-opioid agonists suppress the stimulus effects of cocaine in monkeys.[161] Therefore, it is unlikely that κ-opioid receptors play a direct role in the stimulus of euphoric effects of cocaine. The elevation of the κ_2-opioid receptor subtype along with its discrete localization to the "limbic" or "emotional" striatum indicates that compensatory adaptation in this subtype may underlie the "affective" or "emotional" effects associated with cocaine dependence. Shippenberg and colleagues[162] have suggested that "conditioned aversive effects" associated with hyperactivity of κ-opioidergic neurons in the ventral striatum may underlie the "motivational incentive" to use cocaine. Furthermore, the subjective effects of κ-opioid agonists mimic the symptoms of cocaine withdrawal suggesting that excessive activity of the opioidergic system may, in part, contribute to the aversive effects associated with cocaine withdrawal. Because protracted exposure to cocaine alters the DA-mediated reward systems, κ-opioidergic systems may undergo adaptations in an effort to re-establish the balance between the reward system and the opposing aversive system. However, when cocaine is withdrawn and the dopaminergic reward circuit is no longer activated, the κ_2-opioid receptor numbers may remain elevated, and may contribute to the unpleasant feelings and dysphoria associated with withdrawal from cocaine.

5.4.2 Kappa-Opioid Receptor Drugs as Cocaine Pharmacotherapies

There is considerable evidence supporting a critical role for κ-opioid receptors in the development of cocaine dependence. Co-administration of κ-opioid agonists with cocaine inhibits cocaine self-administration,[163] cocaine-induced place preference,[162] and the development of sensitization to the rewarding effects of cocaine.[93,94,164] In rats, U-69593, a κ-opioid receptor agonist, reduced cocaine self-administration and cocaine seeking behavior[165] and the reinstatement to cocaine self-administration.[166] Further, daily administration of the mixed κ-opioid antagonist and partial μ-opioid agonist buprenorphine reduces cocaine self-administration by rhesus monkeys[167] and prevents the reinstatement of cocaine-reinforced responding in rats.[168] κ-Opioid agonist drugs such as bremazocine also reduce cocaine self-administration in rhesus monkeys.[169,170] These potent anti-cocaine effects exhibited by κ-opioid receptor agents in preclinical animal studies suggest that κ-opioid receptors may be a useful target for the pharmacotherapeutic treatment of cocaine dependence. One κ-opioid agonist, enadoline, has recently been examined in clinical trials. While the drug was well-tolerated it did not reduce the acute subjective effects in a laboratory study.[171] Buprenorphine has demonstrated some efficacy in decreasing cocaine abuse in heroin abusers[172,173] and the combination of buprenorphine and naloxone, a μ-opioid antagonist, is currently used for the treatment of heroin and other opiate addiction.[174] As a medication for cocaine addiction, buprenorphine has been studied in cocaine abusers with concurrent opioid dependence and was found to reduce cocaine use.[175] The development of pharmacotherapeutic κ-opioid agonists has been hindered by reports that, in humans, administration of κ-opioid agonists elicits aversive and psychotomimetic effects.[176–178] The recent identification of multiple subtypes of κ-opioid receptors with distinct pharmacological and molecular properties[140–147] has led to the hypothesis that different κ-opioid receptor subtypes may mediate distinct actions of κ-opioid agonists.[151,152,178] Therefore, it may be possible to develop κ-opioid drugs that lack and/or inhibit the dysphoric properties and yet maintain efficacy for blocking cocaine administration. At present, distinct "sensorimotor" vs. "limbic" striatum may, in part, mediate the feelings of dysphoria and craving associated with cocaine withdrawal distress. The similarity between symptoms associated with cocaine withdrawal and the subjective effects of κ-opioid agonist administration suggests that increased activity in the kappa opioid receptor system during cocaine withdrawal may underlie the dysphoric effects. The extent that the κ_2-opioid receptor subtype specifically mediates the dysphoric properties of κ-opioid agonists will not be known until selective κ_2-opioid agonists are developed. However, if the κ_2-opioid receptor does not mediate dysphoria during cocaine withdrawal, then selective κ_2-opioid receptor antagonists may be useful for the treatment of the dysphoria that underlies relapse and the perpetuation of cocaine misuse.

5.5 SEROTONIN TRANSPORTER AND COCAINE DEPENDENCE

Cocaine binds with high affinity to the serotonin (5-HT) transporter and inhibits 5-HT uptake.[2,179] Serotonergic neurons project from the dorsal raphe to the ventral tegmental area where they modulate mesolimbic DA neurotransmission. Inhibition of 5-HT uptake in the dorsal raphe nucleus by cocaine decreases the firing of the raphe neurons by feedback activation of the 5-HT$_{1A}$ autoreceptors,[180–182] an effect that is blocked by pretreatment with a 5-HT synthesis inhibitor p-chlorophenylalanine.[181] With chronic cocaine treatment, these mechanisms become sensitized probably as a result of a compensatory upregulation of [^{3}H]imipramine binding to the 5-HT transporter in the dorsal raphe, frontal cortex, medial, and sulcal prefrontal cortex of cocaine-treated rats.[182] These studies suggest that adaptations in the serotonergic neurotransmission may contribute in part to the expression of cocaine-induced behaviors. While an enhancement of serotonin neurotransmission is believed to be inhibitory to the expression of cocaine-mediated behaviors or to have minimal effect,[183,184] there is some evidence that 5-HT may play a role in the

mood-elevating effects of acute cocaine. Interestingly, depletion of tryptophan (the precursor to
5-HT) severely attenuated the subjective high experienced by cocaine-dependent subjects.[185] Furthermore, withdrawal from chronic cocaine use has been associated with symptoms of depression[186]
due to cocaine-induced alterations in 5-HT neurotransmission.[187] Together, these studies suggest
that regulatory alterations in serotonergic signaling play a role in cocaine dependence. Furthermore,
drugs that antagonize these alterations in serotonergic systems may be efficacious for the treatment
of cocaine dependence.

5.5.1 The 5-HT Transporter as a Target for Cocaine Pharmacotherapy

The effects of cocaine may be antagonized by 5-HT-mediated inhibition of mesolimbic DA
neurotransmission.[188] Thus, increased 5-HT neurotransmission that results from blocking presynaptic 5-HT uptake may decrease cocaine administration. In keeping, preclinical animal studies have
demonstrated that enhancement of serotonergic neurotransmission by administration of the selective
5-HT uptake inhibitors citalopram and fluoxetine attenuates the discriminative stimulus effects of
cocaine in monkeys.[189] Furthermore, fluoxetine inhibits cocaine self-administration[190] and reduces
the breakpoints on a progressive ration schedule reinforced by cocaine.[191] Conversely, depletion of
5-HT enhances cocaine self-administration.[192] Several 5-HT reuptake inhibitors have been evaluated
for the treatment of cocaine dependence. Fluoxetine significantly decreased subjective ratings of
cocaine's positive mood effects on several visual analog measures and attenuated the mydriatic
effect of cocaine in human cocaine abusers.[193] Fluoxetine has been suggested to decrease craving
and cocaine use in methadone-maintained cocaine users.[194–196] While the efficacy of fluoxetine may
be related to its ability to reduce craving,[197,198] it is likely that its effects are more related to its
ability to reverse the symptoms of depression that are associated with cocaine withdrawal. Another
5-HT transporter inhibitor, sertraline, was not shown to be more effective than placebo in reducing
cocaine use in recent clinical trials.[199]

5.6 GLUTAMATE RECEPTORS AND COCAINE DEPENDENCE

There is increasing evidence supporting a role for glutamate receptors including the NMDA
(N-methyl-D-aspartate) and AMPA receptors in the neural and behavioral changes resulting from
chronic cocaine administration.[7] Glutamate is the major excitatory neurotransmitter found mainly
in cortical and limbic neurons, which project to the nucleus accumbens. Preclinical studies with
the noncompetitive NMDA receptor antagonist MK-801 have linked excitatory glutamatergic synapses with the development of cocaine sensitization, a cardinal feature of cocaine dependence.
Simultaneous administration of low doses of MK-801 prevented the development of sensitization
to the stereotypic and locomotor stimulant effects of cocaine.[200–205] Alternatively, when MK-801
was administered prior to cocaine, the stimulating effects of cocaine were enhanced.[206] The competitive NMDA antagonist CPP partially prevented the development of cocaine sensitization.[204]
MK-801 decreased the incidence of seizures and mortality caused by cocaine.[207–209] The AMPA
receptor antagonist NBQX produced dose-dependent decreases in cocaine-induced locomotor stimulation.[203] Dopaminergic neurons in the ventral tegmental area of cocaine-treated rats were more
responsive to glutamate while nucleus accumbens neurons were less sensitive.[210] Cortical NMDA
receptors are upregulated after cocaine treatment[211] and GluR1 (an AMPA receptor subunit) and
NMDAR1 (an NMDA receptor subunit) are upregulated in the ventral tegmental area,[212] suggesting
that compensatory adaptation of the glutamate receptors may result from or contribute to enhanced
glutamatergic neurotransmission. Alterations in the mesocorticolimbic glutamate transmission may
in part contribute to the development of cocaine sensitization.[210] Recent evidence suggests that
neuroadaptative changes in amygdaloid glutamate receptors, which are involved in cocaine seeking
and craving, are apparent during cocaine withdrawal.[213]

5.6.1 Glutamate Receptors as Targets for Cocaine Pharmacotherapies

Since NMDA receptors mediate the development of sensitization to cocaine's reinforcing effects, they may serve as a target for cocaine pharmacotherapies.[214] However, while both competitive and noncompetitive NMDA receptor antagonists block the development of cocaine sensitization, they appear to be ineffective once sensitization has developed. NMDA receptor antagonists do not alter the acute stimulant effects of cocaine.[7,200] Acute pretreatment with MK-801 caused a loss of discriminative responding; however, it did not block cocaine self-administration.[214] Furthermore, many drugs that act at the NMDA receptors produce phencyclidine-like behavioral effects.[203] Together, these preclinical studies do not offer significant support for NMDA receptor antagonists as cocaine pharmacotherapies. However, AMPA receptor antagonists do not appear to produce phencyclidine-like behavioral effects, and they block cocaine-induced locomotor stimulation. While additional preclinical studies are necessary, it has been suggested that non-NMDA glutamate receptor antagonists, such as agents acting at the metabotropic glutamate receptor 5 (mGluR5), may be a target for the development of pharmacotherapies for the treatment of cocaine dependence.

5.7 GABA RECEPTORS AND COCAINE DEPENDENCE

The GABAergic system in concert with the dopaminergic and glutamatergic systems is involved in cocaine addiction. Dopamine and glutamate terminals synapse on GABA spiny cells in the brain reward area of the nucleus accumbens[215] and there are GABA projections to the nucleus accumbens.[216] There are many similarities in the projections of dopamine and GABA suggesting that GABA may modulate the effects of cocaine within the dopaminergic system. For example, treatment with the dopamine agonist pramipexole was associated with increased GABA levels in the prefrontal cortex of cocaine dependent subject after 8 weeks of treatment.[217] Additionally, acute cocaine use increased dopamine together with increased GABA transmission in the prefrontal cortex,[218] while repeated cocaine use decreased dopamine D_2 receptor and $GABA_B$ receptor function.[219] These changes could ultimately result in lower GABA levels in cocaine-dependent individuals, and, therefore, medications that increase GABA levels may be useful in treating cocaine addiction.

5.7.1 GABA Receptors as Targets for Cocaine Pharmacotherapies

Preclinical and clinical studies have examined the effects of GABA agents on cocaine-seeking behaviors. In animals, gamma-vinyl gamma-aminobutyric acid (GVG), a GABA agonist, reduced cocaine self-administration,[220] and a combination of muscimol, a $GABA_A$ agonist, and baclofen, a $GABA_B$ agonist, blocked the reinstatement of cocaine[221] and decreased cocaine self-administration.[222,223] In humans, the GABA agonists topiramate,[224] tiagabine,[199] and baclofen[225] decreased cocaine use. Baclofen has also been shown to reduce limbic activation in response to cocaine craving.[225] The use of $GABA_B$ agonists as treatments has been slowed by the adverse side effects including sedation and motor impairment. Recently, positive allosteric modulators at the $GABA_B$ receptor have been developed, which do not have intrinsic activity but interact with already present GABA to enhance its effect.[226] Further research is necessary to determine whether GABA receptor agonists, positive allosteric GABA modulators, or possibly a combination will be clinically useful.

5.8 MULTITARGET PHARMACOTHERAPEUTIC AGENTS

Many of the novel pharmacotherapeutic agents currently under development are directed toward a single molecular target related to cocaine or known to be regulated by cocaine. Although this strategy has been somewhat beneficial, the development of an effective treatment for cocaine dependence may require multisite targeting of distinct neuroreceptor populations that are known to modulate the activity of the drug reward circuit. Cocaine interacts with at least three distinct neurochemical systems in the brain including the dopaminergic, serotonergic, and noradrenergic systems. Cocaine enhances the neurotransmission of each of these systems by blocking the pre-synaptic reuptake. Chronic perturbations of monoaminergic neurotransmission that result from protracted use of cocaine may, in turn, alter cholinergic and glutamatergic neurotransmission by indirect actions. The ability of cocaine to alter signaling of multiple neurochemical pathways in the brain suggests that a multitarget pharmacotherapy may be an optimal approach for the treatment of cocaine dependence.

5.8.1   Ibogaine: The Rain Forest Alkaloid

Ibogaine, the principal alkaloid of the African rain forest shrub Tabernanthe iboga (Apocynaceae family), is currently being evaluated as an agent to treat psychostimulant addiction.[227] Anecdotal reports of ibogaine treatments in opiate-dependent or cocaine-dependent humans describe allevia-tion of drug "craving" and physical signs of opiate withdrawal after a single administration of ibogaine, which in some subjects contributes to drug-free periods lasting several months thereafter. This has recently been confirmed in a preliminary study reporting that ibogaine significantly reduced craving for both cocaine and heroin and significantly improved depressive symptoms in an inpatient detoxification setting.[228] In drug self-administration studies, ibogaine and related iboga alkaloids reduced intravenous self-administration of cocaine 1 h after treatment. This suppression on cocaine intake was evident 1 day later, and in some rats a persistent decrease was noted for as long as several weeks.[229] Ibogaine also effectively blocks morphine self-administration[229] and reduces preference for cocaine consumption in a mouse cocaine-preference drinking model.[230] And, cocaine-induced stereotypy and locomotor activity were significantly lower in ibogaine-treated mice.

The mechanism of action for ibogaine may be resolved in part by defining high-affinity pharmacological targets for ibogaine. The receptor binding profile for ibogaine suggests that multiple neurochemical pathways may be responsible for ibogaine's anti-addictive properties. Ibogaine binds to μ- and κ_1-opioid receptors, α-1 adrenergic receptors, M_1 and M_2 muscarinic receptors, serotonin 5-HT_2 and 5-HT_3 receptors, and voltage-dependent sodium channels with micromolar affinities.[231] Ibogaine completely displaced [^{3}H]MK-801 binding[232,233] and blocked NMDA-depolarizations in frog motor neurons.[233] Ibogaine demonstrated moderate affinity for binding to cocaine recognition sites on the DA transporter[231] and on the 5-HT transporter.[234] Ibogaine, which blocks access of cocaine to the DA transporter, may[235] or may not restrict substrate uptake.[236] Because ibogaine displays lower affinity for the DA transporter compared to cocaine, it may meet some of the criteria for the "ideal cocaine antagonist."

The anti-addictive properties of ibogaine may, in part, be mediated by a pharmacologically active metabolite. Recently, the principal metabolite of ibogaine was isolated from biological specimens of subjects administered ibogaine using GC/MS.[234,237] The metabolite that results from O-demethylation of the parent drug was identified as 12-hydroxyibogamine (noribogaine). Prelim-inary pharmacokinetic studies have suggested that noribogaine is generated rapidly and exhibits a slow clearance rate.[234] The relatively long half-life of noribogaine suggests that the long-term biological effects of ibogaine may, in part, be mediated by its metabolite. Similar to ibogaine, noribogaine binds to the μ- and κ_1-opioid receptors with micromolar potency.[238,239] The most striking finding has been the demonstration that noribogaine binds to the cocaine recognition site on the 5-HT transporter with a nanomolar potency,[234,239] and elevates extraneuronal 5-HT in a dose-dependent manner.[234,236] Given the recent evidence that serotonin uptake blockers alleviate some of the symp-toms associated with psychostimulant "craving," these findings suggest that the effects of nori-bogaine on the 5-HT transmission may account, in part, for the potential of ibogaine to interrupt drug-seeking behavior in humans. Overall, it may be suggested that the putative efficacy of ibogaine as a pharmacotherapy for cocaine dependence may be attributed to the combined actions of the parent and the metabolite at multiple CNS targets.[233,239]

5.9 CONCLUSIONS

Significant advances have been made in understanding the neurochemical consequences of cocaine dependence in the past decade. Integration of the findings observed for cocaine's effects on behavior, together with the identification of the receptors and transporters that undergo compensatory adaptations to neutralize cocaine's effects, has led to the identification of several potential neurochemical targets for the development of cocaine pharmacotherapies. Pharmacotherapies that target one or more of the neurochemical systems that have been altered by protracted cocaine use may alleviate the dysphoria, depression, and anxiety that underlie relapse and compulsive cocaine use.

ACKNOWLEDGMENTS

This work was supported by the M.I.R.E.C.C. and the NIMH Biological Sciences Training Program.

REFERENCES

1. Ritz M., Lamb S., Goldberg S., Kuhar M. Cocaine receptors on dopamine transporters are related to self-administration of cocaine. *Science.* 237:1219, 1987.
2. Reith M., Kramer H., Sershen H., Lajtha A. Cocaine completely inhibits catecholamine uptake into brain synaptic vesicles. *Res. Commun. Subst. Abuse.* 10:205, 1989.
3. Kuhar M., Ritz M., Boja J. The dopamine hypothesis of the reinforcing properties of cocaine. *Trends Neurosci.* 14:299, 1991.
4. Mateo Y., Budygin E.A., Morgan D., Roberts D.C., Jones S.R. Fast onset of dopamine uptake inhibition by intravenous cocaine. *Eur. J. Neurosci.* 20:2838, 2004.
5. Gingrich J., Caron M. Recent advances in the molecular biology of dopamine receptors. *Annu. Rev. Neurosci.* 16:299, 1993.
6. Karler R., Calder L., Thai L., Bedingfield J. A dopaminergic-glutamatergic basis for the action of amphetamine and cocaine. *Brain Res.* 8:658, 1994.
7. Trujillo K., Akil H. Excitatory amino acids and drugs of abuse: a role for N-methyl-D-aspartate receptors in drug tolerance, sensitization and physical dependence. *Drug Alcohol Depend.* 38:139, 1995.
8. Chen N., Reith M. Autoregulation and monoamine interactions in the ventral tegmental area in the absence and presence of cocaine: a microdialysis study in freely moving rats. *J. Pharmacol. Exp. Ther.* 271:1597, 1994.
9. DiChiara G., Imperato A. Opposite effects of my and kappa opiate agonists on dopamine release in the nucleus accumbens and in the dorsal caudate of freely moving rats. *J. Pharmacol.* 244:1067, 1988.
10. Spanagel R., Herz A., Shippenberg T. The effects of opioid peptides on dopamine release in the nucleus accumbens: an *in vivo* microdialysis study. *J. Neurochem.* 55:1734, 1990.
11. Spanagel R., Herz A., Shippenberg T. Opposing tonically active endogenous opioid systems modulate the mesolimbic dopaminergic pathway. *Proc. Natl. Acad. Sci. U.S.A.* 89:2046, 1992.
12. Vaughan R.A., Parnas M.L., Gaffaney J.D. et al. Affinity labeling the dopamine transporter ligand binding site. *J. Neurosci. Methods.* 143:33, 2005.
13. Rocha B.A., Fumagalli F., Gainetdinov R.R. et al. Cocaine self-administration in dopamine-transporter knockout mice. *Nat. Neurosci.* 1:132, 1998.
14. Sora I., Hall F.S., Andrews A.M. et al. Molecular mechanisms of cocaine reward: combined dopamine and serotonin transporter knockouts eliminate cocaine place preference. *Proc. Natl. Acad. Sci. U.S.A.* 98:5300, 2001.

15. Alburges M., Narang N., Wamsley J. Alterations in the dopaminergic receptor system after chronic administration of cocaine. *Synapse.* 14:314, 1993.

16. Wilson J., Nobrega J., Carroll M. et al. Heterogenous subregional binding patterns of ^{3}H-WIN 35,428 and ^{3}H-GBR 12,935 are differentially regulated by chronic cocaine self-administration. *J. Neurosci.* 14:2966, 1994.

17. Hitri A., Little K., Ellinwood D. Effect of cocaine on dopamine transporter receptors depends on routes of chronic cocaine administration. *Neuropsychopharmacology.* 14:205, 1996.

18. Aloyo V., Pazalski P., Kirifides A., Harvey J. Behavioral sensitization, behavioral tolerance, and increased [^{3}H]WIN 35,428 binding in rabbit caudate nucleus after repeated injections of cocaine. *Pharmacol. Biochem. Behav.* 52:335, 1995.

19. Letchworth S.R., Nader M.A., Smith H.R., Friedman D.P., Porrino L.J. Progression of changes in dopamine transporter binding site density as a result of cocaine self-administration in rhesus monkeys. *J. Neurosci.* 21:2799, 2001.

20. Little K., McLaughlin D., Zhang L. et al. Brain dopamine transporter messenger RNA and binding sites in cocaine users. *Arch. Gen. Psychiatry.* 55:793, 1998.

21. Little K., Zhang L., Desmond T., Frey K., Dalack G., Cassin B. Striatal dopaminergic abnormalities in human cocaine users. *Am. J. Psychiatry.* 156:238, 1999.

22. Little K., Kirkman J., Carroll F., Clark T., Duncan G. Cocaine use increases [^{3}H]WIN 35, 428 binding sites in human striatum. *Brain Res.* 628:17, 1993.

23. Staley J., Basile M., Wetli C. et al. Differential regulation of the dopamine transporter in cocaine overdose deaths. *Natl. Inst. Drug Abuse Res. Monogr.* 32, 1994.

24. Staley J., Hearn W., Ruttenber A., Wetli C., Mash D. High affinity cocaine recognition sites on the dopamine transporter are elevated in fatal cocaine overdose victims. *Pharmacol. Exp. Ther.* 271:1678, 1994.

25. Staley J., Wetli C., Ruttenber A., Dearn W., Mash D. Altered dopaminergic synaptic markers in cocaine psychosis and sudden death. *Natl. Inst. Drug Abuse Res. Monogr.* 153:491, 1995.

26. Hurd Y., Herkenham M. Molecular alterations in the neostriatum of human cocaine addicts. *Synapse.* 13:357, 1993.

27. Wilson J., Levey A., Bergeron C. et al. Striatal dopamine, dopamine transporter, and vesicular monoamine transporter in chronic cocaine users. *Ann. Neurol.* 40:428, 1996.

28. Kunko P.M., Loeloff R.J., Izenwasser S. Chronic administration of the selective dopamine uptake inhibitor GBR 12,909, but not cocaine, produces marked decreases in dopamine transporter density. *Naunyn Schmiedeberg's Arch. Pharmacol.* 356:562, 1997.

29. Malison R. SPECT imaging of DA transporters in cocaine dependence with [^{123}I]B-CIT. *Natl. Inst. Drug Abuse Res. Monogr.* 152, 1995.

30. Mash D., Pablo J., Ouyang Q., Hearn W., Izenwasser S. Dopamine transport function is elevated in cocaine users. *J. Neurochem.* 81:292, 2002.

31. Daws L., Callaghan P., Morom J. et al. Cocaine increases dopamine uptake and cell surface expression of dopamine transporters. *Biochem. Biophys. Res. Commun.* 290:1545, 2002.

32. Little K., Elmer L., Zhong H., Scheys J., Zhang L. Cocaine induction of dopamine transporter trafficking to the plasma membrane. *Mol. Pharmacol.* 61:436, 2002.

33. Chiat L. Reinforcing and subjective effects of methylphenidate in humans. *Behav. Pharmacol.* 5:281, 1994.

34. Pogun S., Scheffel U., Kuhar M. Cocaine displaces [^{3}H]WIN 35,428 binding to dopamine uptake sites *in vivo*, more rapidly than mazindol or GBR 12,909. *Eur. J. Pharmacol.* 198:203, 1991.

35. Stathis M., Sheffel U., Lever S., Boja J., Carroll F., Kuhar M. Rate of binding of various inhibitors at the dopamine transporter *in vivo*. *Psychopharmacology.* 119:376, 1995.

36. Volkow N., Wang G., Fowler J. et al. Methylphenidate and cocaine have a similar *in vivo* potency to block dopamine transporters in the human brain. *Life Sci.* 65:7, 1999.

37. Volkow N., Ding Y., Fowler J. et al. Is methylphenidate like cocaine? *Arch. Gen. Psychiatry.* 52:456, 1995.

38. Volkow N., Fowler J., Wang G. Imaging studies on the role of dopamine in cocaine reinforcement and addiction in humans. *J. Psychopharmacol.* 13:337, 1999.

39. Tsukada H., Harada N., Nishiyama S., Ohba H., Kakiuchi T. Dose-response and duration effects of acute administrations of cocaine and GBR12909 on dopamine synthesis and transporter in the conscious monkey brain: PET studies combined with microdialysis. *Brain Res.* 860:141, 2000.

40. Rothman R., Glowaw J. A review of the effects of dopaminergic agents on human, animals, and drug-seeking behavior, and its implications for medication development. *Mol. Neurobiol.* 10:1, 1995.
41. Rothman R. High affinity dopamine reuptake inhibitors as potential cocaine antagonists: a strategy for drug development. *Life Sci.* 46:PL17, 1990.
42. Baumann M., Char G., Costa B.D., Rice K., Rothman R. GBR 12909 attenuates cocaine-induced activation of mesolimbic dopamine neurons in the rat. *J. Pharmacol. Exp. Ther.* 271:1216, 1994.
43. Schenk S. Effects of GBR 12909, WIN 35,428 and indatraline on cocaine self-administration and cocaine seeking in rats. *Psychopharmacology* (Berlin). 160:263, 2002.
44. Glowa J., Fantegrossi W., Lewis D., Matecka D., Rice K., Rothman R. Sustained decrease in cocaine-maintained responding in rhesus monkeys with 1-[2]-bis(4-flourophenyl)methoxy[ethyl]-4-(3-hydroxy-3-phenylpropyl)piperazinyldecanoate, a long-acting ester derivative of GBR 12909. *J. Med. Chem.* 39:4689, 1996.
45. Villemagne V., Rothman R., Yokoi F. et al. Doses of GBR 12909 that suppress cocaine self-administration in nonhuman primates substantially occupy dopamine transporters as measured by [^{11}C]WIN 35,428 PET scans. *Synapse.* 32:44, 1999.
46. Howell L., Wilcox K. The dopamine transporter and cocaine medication development: drug self-administration in nonhuman primates. *J. Pharmacol. Exp. Ther.* 298:1, 2001.
47. Dworkin S.I., Lambert P., Sizemore G.M., Carroll F.I., Kuhar M.J. RTI-113 administration reduces cocaine self-administration at high occupancy of dopamine transporter. *Synapse.* 30:49, 1998.
48. Sizemore G.M., Davies H M., Martin T.J., Smith J.E. Effects of 2beta-propanoyl-3beta-(4-tolyl)-tropane (PTT) on the self-administration of cocaine, heroin, and cocaine/heroin combinations in rats. *Drug Alcohol Depend.* 73:259, 2004.
49. Roberts D.C., Jungersmith K.R., Phelan R., Gregg T.M., Davies H.M. Effect of HD-23, a potent long acting cocaine-analog, on cocaine self-administration in rats. *Psychopharmacology* (Berlin). 167:386, 2003.
50. Giros B., Wang Y., Sutter S., McLeskey S., Pfil C., Caron M. Delineation of discrete domains for substrate, cocaine, and tricyclic antidepressant interactions using chimeric dopamine-norepinephrine transporters. *J. Biol. Chem.* 269(23), 15985, 1994.
51. Buck K., Amara S. Chimeric dopamine-norepinephrine transporters delineate structural domains influencing selectivity for catecholamines and 1-methyl-4-phenylpyridinium. *Proc. Natl. Acad. Sci. U.S.A.* 91:12584, 1994.
52. Christenson J. Radioimmunoassay for benzoyl ecogine. U.S. patent 4. 102:979, 1978.
53. Leute R., Bolz G. Nitrogen derivates of benzoyl ecgonine. U.S. patent 3888:866, 1975.
54. Mule S., Jukofshy D., Kogan M., Pace A., De Verebey K. Evaluation of the radioimmunoassay for benzoylecgonine (a cocaine metabolite) in human urine. *Clin. Chem.* 23:796, 1977.
55. Bagasra O., Forman L., Howeedy A., Whittle P. A potential vaccine for cocaine abuse prophylaxis. *Immunopharmacology.* 23:173, 1992.
56. Carrera M., Ashley J., Parsons L., Wirschung P., Koob G. Suppression of psycho-active effects of cocaine by active immunization. *Nature.* 378:727, 1995.
57. Slusher B., Jackson P. A shot in the arm for cocaine addiction. *Nat. Med.* 2:26, 1996.
58. Landry D., Zhao K., Yang G.-Q., Glickman M., Georgiadis T. Antibody-catalyzed degradation of cocaine. *Science.* 259:1899, 1993.
59. Basmadjian G., Singh S., Sastrodjojo B. et al. Generation of polyclonal catalytic antibodies against cocaine using transition state analogs of cocaine conjugated to diphtheria toxoid. *Chem. Pharm. Bull.* 43:1902, 1995.
60. Berkman C., Underiner G., Cahsman J. Synthesis of an immunogenic template for the generation of catalytic antibodies for (–) cocaine hydrolysis. *J. Org. Chem.* 61:5686, 1996.
61. Rocio M., Carrera A., Ashley J. et al. Cocaine vaccines: Antibody protection against relapse in a rat model. *PNAS.* 97:6202, 2000.
62. Rocio M., Carrera A., Ashley J., Wirsching P., Koob G., Janda K. A second-generation vaccine protects against the psychoactive effects of cocaine. *PNAS.* 98:1988, 2001.
63. Mets B., Winger G., Cabrera C. et al. A catalytic antibody against cocaine prevents cocaine's reinforcing and toxic effects in rats. *Proc. Natl. Acad. Sci. U.S.A.* 95:10176, 1998.
64. Self D. Cocaine abuse takes a shot. *Nature.* 378:666, 1995.

65. Kosten T., Rosen M., Bond J. et al. Human therapeutic cocaine vaccine: safety and immunogenicity. *Vaccine*. 20:1196, 2002.

66. Kosten T., Owens S.M. Immunotherapy for the treatment of drug abuse. *Pharmacol. Ther.* 108:76, 2005.

67. Martell B.A., Mitchell E., Poling J., Gonsai K., Kosten T.R. Vaccine pharmacotherapy for the treatment of cocaine dependence. *Biol. Psychiatry.* 58:158, 2005.

68. Robledo P., Maldonado-Lopez R., Koob G. Role of the dopamine receptors in the nucleus accumbens in the rewarding properties of cocaine. *Ann. N.Y. Acad. Sci.* 654:509, 1992.

69. Pulvirenti L., Koob G. Dopamine receptor agonists, partial agonists and psychostimulant addiction. *Trends Pharmacol. Sci.* 15:374, 1994.

70. Ranaldi R., Beninger R. The effects of systemic and intracerebral injections of D_1 and D_2 agonists on brain stimulation reward. *Brain Res.* 651:283, 1994.

71. Caine S., Heinrichs S., Coffin V., Koob G. Effects of the dopamine D-1 antagonist SCH23390 microinjected into the accumbens, amygdala or striatum on cocaine self-administration in the rat. *Brain Res.* 692:47, 1995.

72. Sanchez C.J., Bailie T.M., Wu W.R., Li N., Sorg B.A. Manipulation of dopamine d1-like receptor activation in the rat medial prefrontal cortex alters stress- and cocaine-induced reinstatement of conditioned place preference behavior. *Neuroscience.* 119:497, 2003.

73. Anderson S.M., Bari A.A., Pierce R.C. Administration of the D1-like dopamine receptor antagonist SCH-23390 into the medial nucleus accumbens shell attenuates cocaine priming-induced reinstatement of drug-seeking behavior in rats. *Psychopharmacology* (Berlin). 168:132, 2003.

74. Tella S. Differential blockade of chronic versus acute effects of intravenous cocaine by dopamine receptor antagonists. *Pharmacol. Biochem. Behav.* 48:151, 1994.

75. Shippenberg T., Heidbreder C. Sensitization to the conditioned rewarding effects of cocaine: pharmacological and temporal characteristics. *J. Pharmacol. Exp. Ther.* 273:808, 1995.

76. Hubner C., Moreton J. Effect of selective D_1 and D_2 dopamine antagonists on cocaine self-administration in the rat. *Psychopharmacology.* 105:151, 1991.

77. Woolverton W. Effects of D_1 and D_2 dopamine antagonists on the self-administration of cocaine and piribedil by rhesus monkeys. *Pharmacol. Biochem. Behav.* 24, 1986.

78. Koob G., Le H., Creese I. The D_1 dopamine antagonist SCH 23390 increases cocaine self-administration in the rat. *Neurosci. Lett.* 79:315, 1987.

79. Eglimez Y., Jung M., Lane J., Emmett-Oglesby M. Dopamine release during cocaine self-administration in the rat: effect of SCH 23390. *Brain Res.* 701:142, 1995.

80. Caine S., Koob G. Effects of dopamine D-1 and D-2 antagonists on cocaine self-administration under different schedules of reinforcement in the rat. *J. Pharmacol. Exp. Ther.* 270:209, 1994.

81. Khroyan T.V., Platt D.M., Rowlett J.K., Spealman R.D. Attenuation of relapse to cocaine seeking by dopamine D1 receptor agonists and antagonists in non-human primates. *Psychopharmacology* (Berlin). 168:124, 2003.

82. Alleweireldt A.T., Kirschner K.F., Blake C.B., Neisewander J.L. D1-receptor drugs and cocaine-seeking behavior: investigation of receptor mediation and behavioral disruption in rats. *Psychopharmacology* (Berlin). 168:109, 2003.

83. Meador-Woodruff J., Little K., Damask S., Mansour P., Watson S. Effects of cocaine on dopamine receptor gene expression: A study in the postmortem human brain. *Biol. Psychiatry.* 34:348, 1993.

84. Laurier L., Corrigall C., George S. Dopamine receptor density, sensitivity and mRNA levels are altered following self-administration of cocaine in the rat. *Brain Res.* 634:31, 1994.

85. Zhang D., Zhang L., Lou D.W., Nakabeppu Y., Zhang J., Xu M. The dopamine D1 receptor is a critical mediator for cocaine-induced gene expression. *J. Neurochem.* 82:1453, 2002.

86. Unterwald E.M., Kreek M.J., Cuntapay M. The frequency of cocaine administration impacts cocaine-induced receptor alterations. *Brain Res.* 900:103, 2001.

87. Peris J., Boyson S., Cass W. et al. Persistence of neurochemical changes in dopamine systems after repeated cocaine administration. *J. Pharmacol. Exp. Ther.* 253:35, 1990.

88. Claye L., Akunne H., Davis M., DeMattos S., Soliman K. Behavioral and neurochemical changes in the dopaminergic system after repeated cocaine administration. *Mol. Neurobiol.* 11:55, 1995.

89. Zeigler S., Lipton J., Toga A., Ellison G. Continuous cocaine administration produces persisting changes in the brain neurochemistry and behavior. *Brain Res.* 552:27, 1991.

90. Nader M.A., Daunais J.B., Moore T. et al. Effects of cocaine self-administration on striatal dopamine systems in rhesus monkeys: initial and chronic exposure. *Neuropsychopharmacology.* 27:35, 2002.

91. Henry D., White F. Repeated cocaine administration causes persistent enhancement of D_1 dopamine receptor sensitivity within the rat nucleus accumbens. *J. Pharmacol. Exp. Ther.* 258:882, 1991.

92. Henry D.J., White F.J. The persistence of behavioral sensitization to cocaine parallels enhanced inhibition of nucleus accumbens neurons. *J. Neurosci.* 15:6287, 1995.

93. Shippenberg T., Heidbreder C. Kappa opioid receptor agonists prevent sensitization to the rewarding effects of cocaine. *NIDA Res. Monogr.* 153:456, 1994.

94. Shippenberg T., LeFevour A., Heidbreder C. K-opioid receptor agonists prevent sensitization to the conditioned rewarding effects of cocaine. *J. Pharmacol. Exp. Ther.* 276:545, 1996.

95. Spyraki C., Sealfon S. Regulation of dopamine D2 receptor mRNA expression in the olfactory tubercle by cocaine. *Mol. Brain Res.* 19:313, 1993.

96. Przewlocka B., Lason W. Adaptive changes in the proenkephalin and D_2 dopamine receptor mRNA expression after chronic cocaine in the nucleus accumbens and striatum of the rat. *Neuropsychopharmacology.* 5:464, 1995.

97. Unterwald E., Ho A., Rubenfeld J., Kreek M. Time course of the development of behavioral sensitization and dopamine receptor up-regulation during binge cocaine administration. *J. Pharmacol. Exp. Ther.* 270:1387, 1994.

98. Volkow N., Fowler J., Wolf A. et al. Effects of chronic cocaine abuse on postsynaptic dopamine receptors. *Am. J. Psychiatry.* 147:719, 1990.

99. Worsley J.N., Moszczynska A., Falardeau P. et al. Dopamine D1 receptor protein is elevated in nucleus accumbens of human, chronic methamphetamine users. *Mol. Psychiatry.* 5:664, 2000.

100. Nader M., Mach R. Self-administration of the dopamine D_3 agonist 7-OH-DPAT in rhesus monkeys is modified by prior cocaine exposure. *Psychopharmacology.* 125:13, 1996.

101. Acri J., Carter S., Alling K. et al. Assessment of cocaine-like discriminative stimulus effects of dopamine D_3 receptor ligands. *Eur. J. Pharmacol.* 281:R7, 1995.

102. Mallet P., Beninger R. 7-OH-DPAT produced place conditioning in rats. *Eur. J. Pharmacol.* 261:R5, 1994.

103. Lamas X., Negus S., Nader M., Mello N. Effects of the putative dopamine D_3 receptor agonist 7-OH-DPAT in rhesus monkeys trained to discriminate cocaine from saline. *Psychopharmacology.* 124:306, 1996.

104. Mash D.C., Staley J.K. D3 dopamine and kappa opioid receptor alterations in human brain of cocaine-overdose victims. *Ann. N.Y. Acad. Sci.* 877:507, 1999.

105. Staley J., Mash D. Adaptive increase in D_3 dopamine receptors in the brain reward circuits of human cocaine fatalities. *J. Neurosci.* 16:6100, 1996.

106. Wallace D., Mactutus C., Booze R. Repeated intravenous cocaine administrations: locomotor activity and dopamine D_2/D_3 receptors. *Synapse.* 19, 1996.

107. Neisewander J.L., Fuchs R.A., Tran-Nguyen L.T., Weber S.M., Coffey G.P., Joyce J.N. Increases in dopamine D3 receptor binding in rats receiving a cocaine challenge at various time points after cocaine self-administration: implications for cocaine-seeking behavior. *Neuropsychopharmacology.* 29:1479, 2004.

108. Le Foll B., Frances H., Diaz J., Schwartz J.C., Sokoloff P. Role of the dopamine D3 receptor in reactivity to cocaine-associated cues in mice. *Eur. J. Neurosci.* 15:2016, 2002.

109. Landwehrmeyer B., Mengod G., Palacios J. Differential visualization of dopamine D_2 and D_3 receptor sites in the rat brain. A comparative study using in situ hybridization histochemistry and ligand binding autoradiography. *Eur. J. Neurosci.* 5:145, 1993.

110. Landwehrmeyer B., Mengod G., Palacios J. Dopamine D_3 receptor mRNA and binding site in human brain. *Mol. Brain Res.* 18:187, 1993.

111. Meador-Woodruff J., Little K., Damask S., Watson S. Effects of cocaine on D_3 and D_4 receptor expression in the human striatum. *Biol. Psych.* 38:263, 1995.

112. Le Foll B., Diaz J., Sokoloff P. A single cocaine exposure increases BDNF and D3 receptor expression: implications for drug-conditioning. *Neuroreport.* 16:175, 2005.

113. Cox B., Rosser M., Kozlowski M., Duwe K., Neve R., Neve K. Regulation and functional characterization of a rat recombinant dopamine D_3 receptor. *Synapse.* 21:1, 1995.

114. Fishburn C., Belleli D., David C., Carmon S., Fuchs S. A novel short isoform of the D_3 dopamine receptor generated by alternative splicing in the third cytoplasmic loop. *J. Biol. Chem.* 268:5872, 1993.

115. Sokoloff P., Giros B., Martres M. et al. Localization and function of the D_3 dopamine receptor. *Arzneim. Forsch. Drug Res.* 42:224, 1992.

116. Grundt P., Carlson E.E., Cao J. et al. Novel heterocyclic trans olefin analogues of N-{4-[4-(2,3-dichlorophenyl)piperazin-1-yl]butyl}arylcarboxamides as selective probes with high affinity for the dopamine D_3 receptor. *J. Med. Chem.* 48:839, 2005.

117. DeWit H., Wise R. Blockade of cocaine reinforcement in rats with the dopamine receptor blocker pimozide, but not the noradrenergic blockers phentolamine and phenobenzamine. *Can. J. Psychol.* 31:195, 1977.

118. Roberts D., Vickers G. Atypical neuroleptics increase self-administration of cocaine: an evaluation of a behavioral screen for antipsychotic activity. *Psychopharmacology.* 82:1135, 1984.

119. Wilson M., Schuster C. The effects of chlorpromazine on psychomotor stimulant self-administration in rhesus monkeys. *Psychopharmacologia.* 26:115, 1972.

120. Johansen C., Kandel D., Bonese K. The effect of perphenazine on self-administration behavior. *Pharmacol. Biochem. Behav.* 4:427, 1976.

121. Gawin F., Allen D., Humblestone B. Outpatient treatment of "crack" cocaine smoking with flupentixol decanoate. *Arch. Gen. Psychiatry.* 46:322, 1989.

122. Gawin F. Neuroleptic reduction of cocaine-induced paranoia, but not euphoria? *Psychopharmacology.* 90:142, 1986.

123. Crosby R., Halikas J., Carlson G. Pharmacotherapeutic interventions for cocaine abuse: present practices and future directions. *J. Addict. Dis.* 10:13, 1991.

124. Grabowski J., Rhoades H., Silverman P. et al. Risperidone for the treatment of cocaine dependence: randomized, double-blind trial. *J. Clin. Psychopharmacol.* 20:305, 2000.

125. Roberts D., Ranaldi R. Effect of dopaminergic drugs on cocaine reinforcement. *Clin. Neuropharmacol.* 18:S84, 1995.

126. Malcolm R., Hutto B., Philips J., Ballenger J. Pergolide mesylate treatment of cocaine withdrawal. *J. Clin. Psychiatry.* 52:39, 1991.

127. Dackis C., Golf M., Sweeney D., Byron J., Climko R. Single dose bromocriptine reverses cocaine craving. *Psychiatry Res.* 20:261, 1987.

128. Tennant F., Sagherian A. Double-blind comparison of amantadine and bromocriptine for ambulatory withdrawal from cocaine dependence. *Arch. Intern. Med.* 147:109, 1987.

129. Handelsman L., Chordia P., Escovar I., Marion I., Lowinson J. Amantadine for treatment of cocaine dependence in methadone-maintained patients. *Am. J. Psychiatry.* 145:533, 1988.

130. Self D., Barnhart W., Lehman D., Nestler E. Opposite modulation of cocaine-seeking behavior by D1- and D2-like dopamine receptor agonists. *Science.* 271:1586, 1996.

131. Shoptaw S., Watson D.W., Reiber C. et al. Randomized controlled pilot trial of cabergoline, hydergine and levodopa/carbidopa: Los Angeles Cocaine Rapid Efficacy Screening Trial (CREST). *Addiction.* 100(Suppl. 1):78, 2005.

132. Xi Z.X., Gilbert J.G., Pak A.C., Ashby C.R., Jr., Heidbreder C.A., Gardner E.L. Selective dopamine D3 receptor antagonism by SB-277011A attenuates cocaine reinforcement as assessed by progressive-ratio and variable-cost-variable-payoff fixed-ratio cocaine self-administration in rats. *Eur. J. Neurosci.* 21:3427, 2005.

133. Gilbert J.G., Newman A.H., Gardner E.L. et al. Acute administration of SB-277011A, NGB 2904, or BP 897 inhibits cocaine cue-induced reinstatement of drug-seeking behavior in rats: role of dopamine D_3 receptors. *Synapse.* 57:17, 2005.

134. Xi Z.X., Gilbert J., Campos A.C. et al. Blockade of mesolimbic dopamine D_3 receptors inhibits stress-induced reinstatement of cocaine-seeking in rats. *Psychopharmacology* (Berlin). 176:57, 2004.

135. Le Foll B., Schwartz J.C., Sokoloff P. Dopamine D_3 receptor agents as potential new medications for drug addiction. *Eur. Psychiatry.* 15:140, 2000.

136. Pilla M., Perachon S., Sautel F. et al. Selective inhibition of cocaine-seeking behaviour by a partial dopamine D_3 receptor agonist. *Nature.* 400:371, 1999.

137. Wood M.D., Boyfield I., Nash D.J., Jewitt F.R., Avenell K.Y., Riley G.J. Evidence for antagonist activity of the dopamine D_3 receptor partial agonist, BP 897, at human dopamine D_3 receptor. *Eur. J. Pharmacol.* 407:47, 2000.

138. Gorelick D.A., Gardner E.L., Xi Z.X. Agents in development for the management of cocaine abuse. *Drugs.* 64:1547, 2004.

139. Unterwald E.M. Regulation of opioid receptors by cocaine. *Ann. N.Y. Acad. Sci.* 937:74, 2001.

140. Clark J., Liu L., Price M., Hersh B., Edelson M., Pasternak G. Kappa opiate receptor multiplicity: evidence for two U50-488-sensitive K_1 subtypes and a novel K_3 subtype. *J. Pharmacol. Exp. Ther.* 251:461, 1989.

141. Rothman R., France C., Bykov V. et al. Pharmacological activities of optically pure enantiomers of the K opioid agonist, U50,488 and its *cis* diastereomer: evidence for three K receptor subtypes. *Eur. J. Pharmacol.* 167:345, 1989.

142. Rothman R., Bykov V., Coasta R.D., Jacobsen A., Rice K., Brady L. Evidence for four opioid kappa binding sites in guinea pig brain. Presented at International Narcotics Research Conference (INRC) '89. 9, 1990.

143. Wollemann M., Benhye S., Simon J. The kappa-opioid receptor: evidence for the different subtypes. *Life Sci.* 52:599, 1993.

144. Nishi M., Takeshima H., Fukada K., Kato S., Mori K. cDNA cloning and pharmacological characterization of an opioid receptor with high affinities for kappa-subtype selective ligands. *FEBS Lett.* 330:77, 1993.

145. Pan G., Standifer K., Pasternak G. Cloning and functional characterization through antisense mapping of a K_3-related opioid receptor. *Mol. Pharmacol.* 47:1180, 1995.

146. Raynor K., Kong H., Chen Y. et al. Pharmacological characterization of the cloned kappa-, delta-, and mu-opioid receptors. *Mol. Pharmacol.* 45:330, 1993.

147. Simonin F., Gaveriaux-Ruff C., Befort K. et al. k-Opioid receptor in humans: cDNA and genomic cloning, chromosomal assignment, functional expression, pharmacology, and expression pattern in the central nervous system. *Proc. Natl. Acad. Sci. U.S.A.* 92:7006, 1995.

148. Quirion R., Pilapil C., Magnan J. Localization of kappa opioid receptor binding sites in human forebrain using [³H]U69,593: Comparison with [³H]bremazocine. *Cell Mol. Neurobiol.* 7:303, 1987.

149. Staley J., Rothman R., Partilla J. et al. Cocaine upregulates kappa opioid receptors in human striatum. *Natl. Inst. Drug Abuse Res. Monogr.* 162:234, 1996.

150. Itzhak Y. Differential regulation of brain opioid receptors following repeated cocaine administration to guinea pigs. *Drug Alcohol Depend.* 3:53, 1993.

151. Ni Q., Xu H., Partilla J., Costa B.D., Rice K., Rothman R. Selective labeling of K_2 opioid receptors in rat brain by [125I]IOXY: interactions of opioid peptides and other drugs with multiple K_{2a} binding sites. *Peptides.* 14:1279, 1993.

152. Ni Q., Xu H., Partilla J. et al. Opioid peptide receptor studies. Interaction of opioid peptides and other drugs with four subtypes of the K_2 receptor in guinea pig brain. *Peptides.* 16:1083, 1995.

153. Hammer R. Cocaine alters opiate receptor binding in critical brain reward regions. *Synapse.* 3:55, 1989.

154. Unterwald E., Rubenfeld J., Kreek M. Repeated cocaine administration upregulates kappa and mu but not gamma opioid receptors. *Neuroreport.* 5:1613, 1994.

155. Collins S.L., Kunko P.M., Ladenheim B., Cadet J.L., Carroll F.I., Izenwasser S. Chronic cocaine increases kappa-opioid receptor density: lack of effect by selective dopamine uptake inhibitors. *Synapse.* 45:153, 2002.

156. Spangler R., Bo A., Zhou Y., Maggos C., Yuferov V., Kreek M. Regulation of kappa opioid receptor mRNA in the rat brain by "binge" pattern cocaine administration and correlation with preprodynorphin mRNA. *Mol. Brain Res.* 38:71, 1996.

157. Rosin A., Lindholm S., Franck J., Georgieva J. Downregulation of kappa opioid receptor mRNA levels by chronic ethanol and repetitive cocaine in rat ventral tegmentum and nucleus accumbens. *Neurosci. Lett.* 275:1, 1999.

158. Staley J.K., Rothman R.B., Rice K.C., Partilla J., Mash D.C. Kappa2 opioid receptors in limbic areas of the human brain are upregulated by cocaine in fatal overdose victims. *J. Neurosci.* 17:8225, 1997.

159. Broadbent J., Gaspard T., Dworkin S. Assessment of the discriminative stimulus effects of cocaine in the rat: lack of interaction with opioids. *Pharmacol. Biochem. Behav.* 51:379, 1995.

160. Ukai M., Mori E., Kameyama T. Effects of centrally administered neuropeptides on discriminative stimulus properties of cocaine in the rat. *Pharmacol. Biochem. Behav.* 51:705, 1995.

161. Spealman R., Bergman J. Modulation of the discriminative stimulus effects of cocaine by mu and kappa opioids. *J. Pharmacol. Exp. Ther.* 261:607, 1992.

162. Shippenberg T., Herz A., Spanagel R., Bals-Kubik R., Stein C. Conditioning of opioid reinforcement: Neuroanatomical and neurochemical substrates. *Ann. N.Y. Acad. Sci.* 654:347, 1992.

163. Glick S., Maisonneuve I., Raucci J., Archer S. Kappa opioid inhibition of morphine and cocaine self-administration in rats. *Brain Res.* 681:147, 1995.

164. Heidbreder C., Goldberg S., Shippenberg T. The kappa-opioid receptor agonist U-69,593 attenuates cocaine-induced behavioral sensitization in the rat. *Brain Res.* 616:335, 1993.

165. Schenk S., Partridge B., Shippenberg T.S. U69593, a kappa-opioid agonist, decreases cocaine self-administration and decreases cocaine-produced drug-seeking. *Psychopharmacology* (Berlin). 144:339, 1999.

166. Schenk S., Partridge B., Shippenberg T.S. Reinstatement of extinguished drug-taking behavior in rats: effect of the kappa-opioid receptor agonist, U69593. *Psychopharmacology* (Berlin). 151:85, 2000.

167. Mellow N., Kamein J., Lukas S., Mendelson J., Drieze J., Sholar J. Effects of intermittent buprenorphine administration of cocaine self-administration by rhesus monkeys. *J. Pharmacol. Exp. Ther.* 264:530, 1993.

168. Comer S., Lac S., Curtis L., Carroll M. Effects of buprenorphine and naltrexone on reinstatement of cocaine-reinforced responding in rats. *J. Pharmacol. Exp. Ther.* 267:1470, 1993.

169. Cosgrove K., Carroll M. Effects of bremazocine on self-administration of smoked cocaine base and orally delivered ethanol, phencyclidine, saccharin, and food in rhesus monkeys: a behavioral economic analysis. *J. Pharmacol. Exp. Ther.* 301:Jun 2002, 2002.

170. Mello N.K., Negus S.S. Effects of kappa opioid agonists on cocaine- and food-maintained responding by rhesus monkeys. *J. Pharmacol. Exp. Ther.* 286:812, 1998.

171. Walsh S.L., Geter-Douglas B., Strain E.C., Bigelow G.E. Enadoline and butorphanol: evaluation of kappa-agonists on cocaine pharmacodynamics and cocaine self-administration in humans. *J. Pharmacol. Exp. Ther.* 299:147, 2001.

172. Kosten T. Pharmacological approaches to cocaine dependence. *Clin. Neuropharmacol.* 15(Suppl. 70A), 1992.

173. Fudala P., Johnson R., Jaffe J. Outpatient comparison of buprenorphine and methadone maintenance. II. Effects of cocaine usage, retention time in study and missed clinical visits. In Harrison L., Ed. Problems of Drug Dependence. *Natl. Inst. Mental Health Res. Monogr.* 105:587, 1991.

174. Wesson D.R. Buprenorphine in the treatment of opiate dependence: its pharmacology and social context of use in the U.S. *J. Psychoactive Drugs.* Suppl. 2:119, 2004.

175. Montoya I.D., Gorelick D.A., Preston K.L. et al. Randomized trial of buprenorphine for treatment of concurrent opiate and cocaine dependence. *Clin. Pharmacol. Ther.* 75:34, 2004.

176. Pfeiffer A., Brandt V., Herz A. Psychotomimesis mediated by kappa opiate receptors. *Science.* 233:774, 1986.

177. Kumor K., Haertzen C., Johnson R., Kocher T., Jasinski D. Human psychopharmacology of ketocyclazocine as compared with cyclazocine, morphine and placebo. *J. Pharmacol. Exp. Ther.* 238:960, 1986.

178. Herz A. Implications of the multiplicity of opioid receptors for the problem of addiction. *Drug Alcohol Depend.* 25:125, 1990.

179. Koe B. Molecular geometry of inhibitors of the uptake of catecholamines and serotonin in synaptosomal preparations of rat brain. *J. Pharmacol. Exp. Ther.* 199:649, 1976.

180. Pitts D., Marwah J. Cocaine modulation of central momoaminergic neurotransmission. *Pharmacol. Biochem. Behav.* 26:453, 1987.

181. Cunningham K., Lakoski J. The interaction of cocaine with serotonin dural raphe neurons. Single unit extracellular recording studies. *Neuropsychopharmacology.* 3:41, 1990.

182. Cunningham K., Paris J., Goeders N. Chronic cocaine enhances serotonin autoregulation and serotonin uptake binding. *Synapse.* 11:112, 1992.

183. Reith M., Fischette C. Sertraline and cocaine-inducted locomotion in mice. II. Chronic studies. *Psychopharmacology.* 103:306, 1991.

184. Reith M., Wiener H., Fischette C. Sertraline and cocaine-induced locomotion in mice. I. Acute studies. *Psychopharmacology.* 103:306, 1991.

185. Aronson S., Black J., McDougle C. et al. Serotonergic mechanisms of cocaine effects in humans. *Psychopharmacology.* 119:179, 1995.

186. Zeidonis D., Kosten T. Depression as a prognostic factor for pharmacological treatment of cocaine dependence. *Psychopharmacol. Bull.* 27:337, 1991.

187. Parsons L., Koob G., Weiss F. Serotonin dysfunction in the nucleus accumbens of rats during withdrawal after unlimited access to intravenous cocaine. *J. Pharmacol. Exp. Ther.* 274:1182, 1995.

188. Galloway M. Regulation of dopamine and serotonin synthesis by acute administration of cocaine. *Synapse.* 6:63, 1990.

189. Spealman R. Modification of behavioral effect of cocaine by selective serotonin and dopamine uptake inhibitors in squirrel monkeys. *Psychopharmacology.* 112:93, 1993.

190. Carroll M., Lac S., Asencio M., Kragh R. Fluoxetine reduces intravenous cocaine self-administration in rats. *Pharmacol. Biochem. Behav.* 35:237, 1990.

191. Richardson N., Roberts D. Fluoxetine pre-treatment reduced breaking points on a progressive ratio schedule reinforced by intravenous cocaine administration in the rat. *Life Sci.* 49:833, 1991.

192. Lyness W., Friedle N., Moore K. Increased self-administration of d-amphetamine, self-administration. *Pharmacol. Biochem. Behav.* 12:937, 1980.

193. Walsh S., Preston K., Sullivan J., Fromme R., Bigelow G. Fluoxetine alters the effects of intravenous cocaine in humans. *J. Clin. Psychopharmacol.* 14:396, 1994.

194. Batki S., Washburn A., Manfredi L. et al. Fluoxetine in primary and secondary cocaine dependence: outcome using quantitative benzoylecgonine concentration. *Natl. Inst. Drug Abuse Res. Monogr.* 141:140, 1994.

195. Batki S., Manfredi L., Jacob P., Jones R. Fluoxetine for cocaine dependence in methadone maintenance: quantitative plasma and urine cocaine/benzoylecgonine concentrations. *J. Clin. Psychopharmacol.* 13:243, 1993.

196. Pollack M., Rosenbaum J. Fluoxetine treatment of cocaine abuse in heroin addicts. *J. Clin. Psychiatry.* 52:31, 1991.

197. Satel S. Craving for and fear of cocaine: A phenomenologic update on cocaine craving and paranoia. In Kosten T., Kleber H., Eds. *Clinician's Guide to Cocaine Addiction.* Guilford Press, New York, 1992, 172.

198. Satel S., Krystal J., Delgado P., Kosten T., Charney D. Tryptophan depletion and attenuation of cue-induced craving for cocaine. *Am. J. Psychiatry.* 152:778, 1995.

199. Winhusen T.M., Somoza E.C., Harrer J.M. et al. A placebo-controlled screening trial of tiagabine, sertraline and donepezil as cocaine dependence treatments. *Addiction.* 100(Suppl. 1):68, 2005.

200. Karler R., Calder L., Chaudhry I., Turkanis S. Blockade of 'reverse tolerance' to cocaine and amphetamine by MK-801. *Life Sci.* 45:599, 1989.

201. Pudiak C., Bozarth M. L-NAME and MK-801 attenuate sensitization to the locomotor-stimulating effect of cocaine. *Life Sci.* 53:1517, 1993.

202. Wolf M., Jeziorski M. Coadministration of MK-801 with amphetamine, cocaine or morphine prevents rather than transiently masks the development of behavioral sensitization. *Brain Res.* 613:291, 1993.

203. Witkin J. Blockade of the locomotor stimulant effects of cocaine and methamphetamine by glutamate antagonists. *Life Sci.* 53:PL 405, 1993.

204. Haracz J., Belanger S., MacDonall J., Sircar R. Antagonists of *N*-methyl-D-aspartate receptors partially prevent the development of cocaine sensitization. *Life Sci.* 57:2347, 1995.

205. Ida I., Aami T., Kuribara H. Inhibition of cocaine sensitization by MK-801 a noncompetitive *N*-methyl-D-aspartate (NMDA) receptor antagonist: evaluation by ambulatory activity in mice. *Jpn. J. Pharmacol.* 69:83, 1995.

206. Carey R., Dai H., Krost M., Huston J. The NMDA Receptor and cocaine: evidence that MK-801 can induce behavioral sensitization effects. *Pharmacol. Biochem. Behav.* 51:901, 1995.

207. Rockhold R., Oden G., Ho I., Andrew M., Farley J. Glutamate receptor antagonists block cocaine-induced convulsions and death. *Brain Res. Bull.* 27:721, 1991.

208. Itzhak Y., Stein I. Sensitization to the toxic effects of cocaine in mice is associated with regulation of *N*-methyl-D-aspartate receptors in the cortex. *J. Pharmacol. Exp. Ther.* 262:464, 1992.

209. Shimosato K., Marley R., Saito T. Differential effects of NMDA receptor and dopamine receptor antagonists on cocaine toxicities. *Pharmacol. Biochem. Behav.* 51, 1995.

210. White F., Hu X., Zhang X., Wolf M. Repeated administration of cocaine or amphetamines alters neuronal responses to glutamate in the mesoaccumbens dopamine system. *J. Pharmacol. Exp. Ther.* 273:445, 1995.

211. Itzhak Y. Modulation of the PCP/NMDA receptor complex and sigma binding by psychostimulants. *Neurotoxicol. Teratol.* 16:363, 1994.

212. Fitzgerald L., Oritz J., Hamedani A., Nestler E. Drugs of abuse and stress increase the expression of GluR1 and NMDAR1 glutamate receptor subunits in the rat ventral tegmental area: common adaptations among cross-sensitizing agents. *J. Neurosci.* 16:274, 1996.

213. Lu L., Dempsey J., Shaham Y., Hope B.T. Differential long-term neuroadaptations of glutamate receptors in the basolateral and central amygdala after withdrawal from cocaine self-administration in rats. *J. Neurochem.* 94:161, 2005.

214. Schenk S., Valadez A., McNamara C. et al. Development and expression of sensitization to cocaine's reinforcing properties: role of NMDA receptors. *Psychopharmacology.* 111:332, 1993.

215. Sesack S.R., Pickel V.M. In the rat medial nucleus accumbens, hippocampal and catecholaminergic terminals converge on spiny neurons and are in apposition to each other. *Brain Res.* 527:266, 1990.

216. Pennartz C.M., Groenewegen H.J., Lopes da Silva F.H. The nucleus accumbens as a complex of functionally distinct neuronal ensembles: an integration of behavioural, electrophysiological and anatomical data. *Prog. Neurobiol.* 42:719, 1994.

217. Streeter C.C., Hennen J., Ke Y. et al. Prefrontal GABA levels in cocaine-dependent subjects increase with pramipexole and venlafaxine treatment. *Psychopharmacology* (Berlin). 1, 2005.

218. Dackis C.A., O'Brien C.P. Cocaine dependence: a disease of the brain's reward centers. *J. Subst. Abuse Treat.* 21:111, 2001.

219. Jayaram P., Steketee J.D. Effects of cocaine-induced behavioural sensitization on GABA transmission within rat medial prefrontal cortex. *Eur. J. Neurosci.* 21:2035, 2005.

220. Stromberg M.F., Mackler S.A., Volpicelli J.R., O'Brien C.P., Dewey S.L. The effect of gamma-vinyl-GABA on the consumption of concurrently available oral cocaine and ethanol in the rat. *Pharmacol. Biochem. Behav.* 68:291, 2001.

221. McFarland K., Kalivas P.W. The circuitry mediating cocaine-induced reinstatement of drug-seeking behavior. *J. Neurosci.* 21:8655, 2001.

222. Campbell U.C., Lac S.T., Carroll M.E. Effects of baclofen on maintenance and reinstatement of intravenous cocaine self-administration in rats. *Psychopharmacology* (Berlin). 143:209, 1999.

223. Roberts D.C., Andrews M.M., Vickers G.J. Baclofen attenuates the reinforcing effects of cocaine in rats. *Neuropsychopharmacology.* 15:417, 1996.

224. Kampman K.M., Pettinati H., Lynch K.G. et al. A pilot trial of topiramate for the treatment of cocaine dependence. *Drug Alcohol Depend.* 75:233, 2004.

225. Brebner K., Childress A.R., Roberts D.C. A potential role for GABA(B) agonists in the treatment of psychostimulant addiction. *Alcohol Alcohol.* 37:478, 2002.

226. Roberts D.C. Preclinical evidence for GABA(B) agonists as a pharmacotherapy for cocaine addiction. *Physiol. Behav.* 86:18, 2005.

227. Sanchez-Ramos J., Mash D. Ibogaine Human Phase I Pharmacokinetic and Safety Trial. FDA IND. 3968, 1993 (revised 1995).

228. Mash D.C., Kovera C.A., Pablo J. et al. Ibogaine: complex pharmacokinetics, concerns for safety, and preliminary efficacy measures. *Ann. N.Y. Acad. Sci.* 914:394, 2000.

229. Glick S., Kuehne M., Caucci J. et al. Effects of iboga alkaloids on morphine and cocaine self-administration in rats: relationship to tremorigenic effects and to effects on dopamine release in nucleus accumbens and striatum. *Brain Res.* 657:14, 1994.

230. Sershen H., Hashim A., Lajtha A. Ibogaine reduces preference for cocaine consumption in C57BL/6By mice. *Pharmacol. Biochem. Behav.* 47:13, 1994.

231. Sweetnam P., Lancaster J., Snowman A. et al. Receptor binding profile suggests multiple mechanisms of action are responsible for ibogaine's putative anti-addictive activity. *Psychopharmacology.* 118:369, 1995.

232. Popik P., Layer R., Skolnik P. The putative anti-addictive drug ibogaine is a competitive inhibitor of [^{3}H]MK-801 binding to the NMDA receptor complex. *Psychopharmacology.* 114:672, 1994.

233. Mash D., Staley J., Pablo J., Holohean A., Hackman J., Davidoff R. Properties of ibogaine and its principal metabolite (12-hydroxyibogamine) at the MK-801 binding site of the NMDA complex. *Neurosci. Lett.* 192:53, 1995.

234. Mash D., Staley J., Baumann M., Rothman R., Hearn W. Identification of a primary metabolite of ibogaine that targets serotonin transporters and elevates serotonin. *Life Sci.* 57:PL45, 1995.

235. Baumann M. Personal communication.

236. Broderick P., Phelan F., Eng F., Wechsler R. Ibogaine modulates cocaine responses which are altered due to environmental habituation: *In vivo* microvoltammetric and behavioral studies. *Pharmacol. Biochem. Behav.* 49:711, 1994.
237. Hearn W., Pablo J., Hime G., Mash D. Identification and quantitation of ibogaine and an *O*-demethylated metabolite in brain and biological fluids using gas chromatography mass spectrometry. *J. Anal. Toxicol.* 19:427, 1995.
238. Pearl S., Herrick-Davis K., Teitler M., Glick S. Radioligand binding study of noribogaine a likely metabolite of ibogaine. *Brain Res.* 675:342, 1995.
239. Staley J., Ouyang Q., Pablo J. et al. Pharmacological screen for activities of 12-hydroxyibogamine: a primary metabolite of the indole alkaloid ibogaine. *Psychopharmacology.* 127:10, 1996.

CHAPTER 6

Neuropsychiatric Consequences of Chronic Cocaine Abuse

Deborah C. Mash, Ph.D.

Departments of Neurology and Molecular and Cellular Pharmacology, University of Miami, Miller School of Medicine, Miami, Florida

CONTENTS

Mortality data have indicated that deaths involving psychostimulant drugs stem not only from overdose, but also from drug-induced mental states that may lead to serious injuries.[1] The arrival of inexpensive smokable "crack" cocaine has radically changed the nature of the epidemic and revealed the great addictive potential of cocaine. Cocaine, particularly smoked "crack" cocaine, is known to be one of the most widely abused psychoactive substances in the U.S. With the increased use of cocaine in its various forms over the past 15 years, researchers and clinicians have focused on the definition of cocaine dependence and withdrawal.[2] Cocaine was not thought to be addictive prior to the 1980s, as neither chronic use nor its cessation resulted in the physiological tolerance or withdrawal observed in opiate dependence. The progression of occasional use to compulsive use,[3] and the description of a cocaine abstinence syndrome,[4] has led to the definition of diagnostic criteria for cocaine dependence. Clinical experience has fostered the view that persons with psychiatric disorders tend to have high rates of substance abuse, and vice versa.[5,6] Epidemiological studies demonstrate that a large portion of the population experiences both mental and addictive disorders.[7] These studies have underscored the gravity of the problem of dual diagnoses of mental health and substance abuse disorders.

Table 6.1   Behavioral Signs of Acute Psychostimulant Toxicity

Excitability
Restlessness
Delusions
Hallucinations
Paranoia
Panic Attacks
Agitated Delirium

6.1 DIFFERENTIAL DIAGNOSIS OF PSYCHOTIC DISORDERS

Drug use is a major complicating factor in psychosis; it renders the management of psychotic disorders more difficult, and adverse reactions to recreational drugs may mimic psychosis.[8] The differential diagnosis of psychotic disorders in the young routinely includes "drug induced psychosis." This diagnostic category has not had consistent definition and the relationship between drug use and psychotic symptoms is controversial. Adverse psychiatric effects associated with acute cocaine intoxication include extreme agitation, irritability or affective liability, impaired judgment, paranoia, hallucinations (visual or tactile), and, sometimes, manic excitement. Medical and psychiatric symptoms caused by acute cocaine intoxication are a common reason for presentation to the emergency department. Psychiatric symptoms of cocaine intoxication usually subside within 24 h, but some patients may require benzodiazepines for acute agitation. Neuroleptics are often used for the treatment of unremitting paranoid psychosis, hallucinations, and delusions. The transient paranoid state is a common feature of cocaine dependence, with affected persons possessing an obvious predisposition to this drug-induced state.[9] Psychiatric complications of cocaine intoxication include cocaine-induced paranoia, agitated delirium, delusional disorder, and the depressed mood and dysphoria associated with abrupt cocaine withdrawal (Table 6.1).

Extended behavioral signs of cocaine psychosis usually imply the presence of an underlying major psychopathology in susceptible individuals.[9] Cocaine-induced psychosis typically manifests as an intense hypervigilance (paranoia) accompanied by marked apprehension and fear. Auditory and tactile hallucinations, formal thought disorder, and ideas of reference frequently noted with chronic use of amphetamines are not prevalent in cocaine abusers. Paranoid experience secondary to cocaine use is usually limited to a drug episode, which dissipates by the time the user awakens from the "crash," usually about 8 to 36 h after the cessation of the cocaine "binge."[10] In a sample of 100 cocaine-dependent males, none reported cocaine paranoia extending beyond the crash phase.[10]

In contrast to the effects of cocaine, amphetamine has greater and longer-acting psychotogenic properties.[11] Angrist[11] has suggested that high rates of cocaine use that cause sustained elevations in plasma levels may be necessary for the development or kindling of an episode of cocaine psychosis. In keeping with this suggestion, certain cocaine-induced effects are known to become progressively more intense after repeated administration, a phenomenon referred to as sensitization. However, Satel and co-workers[10] have provided data to suggest that instances of cocaine-induced paranoia or psychosis lasting more than several days most likely indicate the presence of an underlying primary psychotic disorder.

6.2 COCAINE DELIRIUM

Delirium symptoms suggest dysfunction of multiple brain regions.[12] Clinical subtypes of delirium with unique and definable phenomenological or physical characteristics are not widely accepted. At present, very little information is known about the neuropathogenesis of cocaine

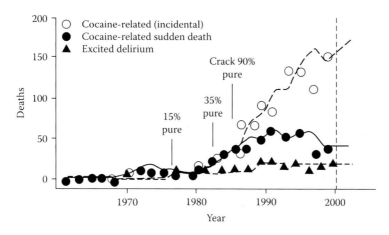

Figure 6.1   Tracking the incidence of cocaine overdose deaths in Dade County, FL. Medicolegal investigations of the deaths were conducted by forensic pathologists. Forensic pathologists evaluated the scene environment and circumstances of death and autopsied the victim in order to determine the cause and manner of death. The circumstances of death and toxicology results were reviewed before classifying a death due to cocaine toxicity with or without preterminal delirium. There was a sharp increase in the incidence of cocaine-related and cocaine overdose cases with the arrival of "crack" cocaine in Dade County. The incidence of cocaine delirium victims is shown by year, from the first report in 1982.

delirium. While various neurotransmitter alterations may converge to result in a delirium syndrome or subtype thereof, an excess of the neurotransmitter dopamine (DA) has been implicated as a cause of cocaine delirium.

In 1985, a case series of cocaine overdose victims who died following a syndrome of excited delirium was first described.[13] It was not clear whether this type of cocaine toxicity represented a new syndrome that was associated with cocaine use alone, or whether there were other causes or underlying genetic risk factors. The cocaine delirium syndrome comprises four components that appear in sequence: hyperthermia, delirium with agitation, respiratory arrest, and death. An episode of cocaine delirium is most often seen at the end of one or more days of drug use.[14] Compared with other accidental cocaine toxicity deaths, a larger proportion of victims of cocaine delirium survive longer after the onset of the overdose. This factor probably accounts for the lower blood cocaine concentrations reported for cocaine delirium victims.[15] The incidence of this disorder is not known with any certainty, but the number of cases has increased markedly since the beginning of the epidemic of "crack" cocaine use in Dade County, FL (Figure 6.1).

In the original report of Wetli and Fishbain,[13] they described the cocaine delirium syndrome in seven cases, and all had somewhat stereotyped histories. A typical example of a cocaine delirium victim was the case of a 33-year-old man, who in an agitated state started pounding on the door of his former house. He was shouting that he wanted to see his wife and daughter. The occupants informed him that nobody by that name resided there; yet he continued. Four bystanders finally restrained him and assisted police units upon their arrival. The subject was handcuffed and put into a police car, whereupon he began to kick out the windows of the vehicle. The police subsequently restrained his ankles and attached the ankle restraints and handcuffs together. He was then transported to a local hospital. While en route, the police officers noted that he became tranquil. Upon arrival at the hospital approximately 45 min after the onset of the agitated delirium, the subject was discovered to be in a respiratory arrest.

A post-mortem examination and a rectal temperature of 41°C (106°F) were recorded. He had needle marks typical of intravenous drug abuse and pulmonary and cerebral edema. Abrasions and contusions of the ankles and wrists were evident from his struggling. Lidocaine was not administered to the victim during the resuscitative attempts. The clinical presentation of cocaine delirium

**Table 6.2   Common Traits Associated with the
                  Fatal Cocaine Delirium Syndrome**

Male
Extreme agitation
Hyperthermia (>103°F)
High body mass index
Survive longer than 1 h after the onset of symptoms
Die in police custody

is different from that of nonpsychotic cocaine abusers with sudden death or massive drug overdose. The cocaine delirium victims are almost always men, they are more likely to die in custody, and are more likely to survive for more than 1 h after the onset of symptoms (Table 6.2).

In the epidemiological tracking of agitated delirium victims in Metropolitan Dade County, men with preterminal delirium comprised approximately 10% of the annual number of cocaine overdose deaths. The demographic trends show that the proportion of these cases remains consistent throughout the epidemic of cocaine abuse and tends to track the annual frequency of cocaine-related sudden deaths. This observation suggests that a certain percentage of cocaine addicts may be at risk for cocaine delirium with chronic abuse.

Cocaine delirium deaths are seasonal and tend to cluster during the late summer months. Core body temperatures are markedly elevated, ranging from 104°C to 108°C. Based on a review of the constellation of psychiatric symptoms associated with this disorder, Kosten and Kleber[16] have termed agitated delirium as a possible cocaine variant of neuroleptic malignant syndrome. Neuroleptic malignant syndrome (NMS) is a highly lethal disorder seen in patients taking dopamine (DA) antagonists or following abrupt withdrawal from DAergic agonists.[17,18] NMS is usually associated with muscle rigidity, while the cocaine variant of the syndrome presents with brief onset of rigidity immediately prior to respiratory collapse.[19]

At present it is not clear whether extreme agitation, delirium, hyperthermia, and rhabdomyolysis are effects of cocaine that occur independently and at random among cocaine users, or whether these features are linked by common toxicologic and pathologic processes.[20] Ruttenber and colleagues[20] have examined excited delirium deaths in a population-based registry of all cocaine-related deaths in Dade County. This study has led to clear description of the cocaine delirium syndrome, its pattern of occurrence in cocaine users over time, and has identified a number of important risk factors for the syndrome.

Cocaine delirium deaths are defined as accidental cocaine toxicity deaths that occurred in individuals who experienced an episode of bizarre behavior prior to death. Bizarre behavior is defined as hyperactivity accompanied by incoherent shouting, aggression (fighting with others or destroying property), or evidence of extreme paranoia as described by witnesses and supported by scene evidence. The results of this study demonstrate that victims are more likely to be male, black, and younger than other cocaine overdose toxicity deaths. The most frequent route of administration was injection for the excited delirium victims as compared to inhalation for the other accidental cocaine toxicity deaths. The frequency of smoked "crack" cocaine was similar for both groups. Of the excited delirium victims, 39% died in police custody as compared with only 2% for the comparison group of accidental cocaine toxicity cases.[20] A large proportion of these individuals survive between 1 and 12 h after the onset of the syndrome.

The most striking feature of the excited delirium syndrome is the extreme hyperthermia. The epidemiological data[20] provide some clues for the etiology of the elevated body temperature. Victims of cocaine excited delirium have higher body mass indices. This finding suggests that muscle mass and adiposity may contribute to the generation of body heat. Temporal clustering in summer months[13] supports the hypothesis that abnormal thermoregulation is an important risk factor for death in people who develop the syndrome. Being placed in police custody prior to death can also raise body temperature through increased psychomotor activity if the victim struggles in the process of restraint. Descriptions of the circumstances around death suggest that police officers frequently had to forcibly restrain these victims. Positional asphyxia and a restraint-induced increase in catecholamines have been hypothesized as contributing causes of cocaine delirium.[21]

6.3 NEUROCHEMICAL PATHOLOGY OF COCAINE DELIRIUM

The mesolimbic dopaminergic (DAergic) system is an important pathway mediating reinforcement and addiction to cocaine and other psychostimulants.[22] Cocaine potentiates DAergic neurotransmission by binding to the DA transporter and blocking neurotransmitter uptake, leading to marked elevations in synaptic DA (for review, see Reference 23). Long-term cocaine abuse leads to neuroadaptive changes in the signaling proteins that regulate DA homeostasis. DA transporter binding site densities have been shown to be upregulated *in vitro* in the post-mortem brain of cocaine addicts,[24–27] and *in vivo* in acutely abstinent cocaine-dependent individuals.[28]

A number of different studies point to a possibility of a defective interaction of cocaine with the DA transporter in the etiology of cocaine delirium. The effects of chronic, intermittent cocaine treatment paradigms on the labeling of the cocaine recognition sites on the DA transporter have been investigated in rat studies. Neuroadaptive changes in the DA transporter have been characterized with a number of different radioligands, including [^{3}H]cocaine, the cocaine congeners [^{3}H]WIN 35,428 and [^{125}I]RTI-55, and more recently with [^{125}I]RTI-121 (for review, see Reference 29). In contrast to the classic DA transport inhibitors ([^{3}H]mazindol, [^{3}H]GBR 12935, and [^{3}H]nomifensine), the cocaine congeners ([^{3}H]WIN 35,428, [^{125}I]RTI-55, and [^{125}I]RTI-121) label multiple sites with a pharmacological profile characteristic of the DA transporter in rat, primate, and human brain.[30–32] Chronic treatment of rats with intermittent doses of cocaine demonstrated a twofold to fivefold increase in the apparent density of [^{3}H]cocaine binding sites in the striatum.[33] Rats that were allowed to self-administer cocaine in a chronic unlimited access paradigm had significant increases in [^{3}H]WIN 35,428 binding sites when the animals were sacrificed on the last day of cocaine access.[34] Rabbits treated with cocaine (4 mg/kg i.v. 2 × per day for 22 days) show an elevation in the density of [^{3}H]WIN 35,428 binding sites in the caudate.[35] A progression of changes were observed in cocaine self-administering monkeys, which had marked elevations in DA transporter binding sites in the more limbic sectors of the striatum (ventromedial putamen and nucleus accumbens) in monkeys exposed to cocaine for 3 months to 1 year.[36] Taken together, these results demonstrate that cocaine exposure leads to an increase in the density of cocaine binding sites on the DA transport carrier.

Cocaine congeners label high- and low-affinity sites on the cloned and native human DA transporter, one of which appears to overlap with the functional state of the carrier protein.[37] In cocaine overdose victims, high-affinity cocaine recognition sites on the DA transporter were upregulated significantly in the striatum as compared to age-matched and drug-free control subjects (Figure 6.2). If this regulatory change in high affinity [^{3}H]WIN 35,428 binding sites on the human DA transporter reflects an increased ability of the protein to transport DA, it may help to explain the addictive liability of cocaine. In synaptosomes isolated from cryoprotected brain specimens, DA uptake function was elevated twofold in the ventral striatum from cocaine users as compared to age-matched drug-free control subjects.[27] In contrast, the levels of [^{3}H]DA uptake were not elevated in victims of excited cocaine delirium, who experienced paranoia and marked agitation prior to death. In keeping with the increase in DA transporter function, radioligand binding to the DA transporter was increased in the cocaine users, but not in the victims of excited delirium. These results demonstrate that long-term cocaine abuse leads to neuroadaptive changes in the signaling proteins that regulate dopamine homeostasis, including elevated DA transporter function and binding sites.

Since cocaine potentiates dopaminergic neurotransmission by binding to DA transporter and blocking reuptake, persisting increases in DA transporter function after cocaine levels have fallen

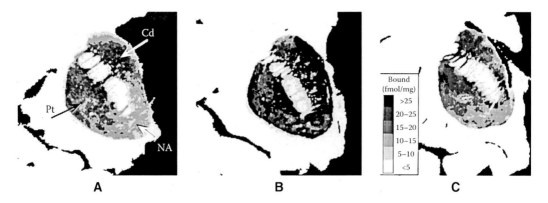

Figure 6.2   *In vitro* autoradiographic maps of [³H]WIN 35,428 labeling of the DA transporter in coronal sections
of the striatum. (A) Representative age-matched and drug-free subject, (B) cocaine overdose
victim, and (C) cocaine delirium victim. The brain maps illustrate the adaptive increase in DA
transporter density over the striatum in the cocaine overdose victim. Note the lack of any apparent
elevation for the victim presenting with agitated delirium. Since the DA transporter regulates the
synaptic concentration of neurotransmitter, the lack of a compensatory upregulation may result in
a DA overflow following a cocaine "binge." Elevated synaptic DA with repeat exposures may kindle
the emergence of the agitated delirium syndrome. Gray scale codes are shown in panel B (black
= high densities; gray = intermediate; light gray to white = low densities). Abbreviations: Cd,
caudate; NA, nucleus accumbens; Pt, putamen.

in blood and brain may result in an acute decrease in the intrasynaptic concentration of DA and
lower DAergic tone. As the transporter carrier upregulates its apparent density in the nerve terminal
to more efficiently transport DA back into the presynaptic nerve terminal, more cocaine will be
needed to experience cocaine's reinforcing effects and euphoria. During acute abstinence from
cocaine, enhanced function of the DA transporter could lead to net depletion in synaptic DA. This
depletion of DA may serve as a biological substrate for anhedonia, the cardinal feature of cocaine
withdrawal symptomatology.

Unlike the results seen in cases of accidental cocaine overdose,[24] the density of high-affinity
cocaine recognition sites on the DA transporter measured in the striatum from cocaine delirium
victims fails to demonstrate a compensatory increase with chronic abuse.[27] Since the concentration
of synaptic DA is controlled by the reuptake mechanism(s), the lack of compensatory increase in
cocaine recognition sites could be the defect in DAergic transmission that explains the paranoia
and agitation associated with this syndrome. Paranoia in the context of cocaine abuse is common
and several lines of evidence suggest that this phenomenon may be related to the function of the
DA transporter protein.[38] Genetic differences in the makeup of individuals who abuse cocaine may
also underlie some of these differences in susceptibility to the development of adverse neuropsychiatric effects with chronic cocaine abuse, that appear to result from a defective regulation of the
DA transporter protein.[38] In addition to the adverse neuropsychiatric sequalae, cocaine delirium
victims are distinguished from other accidental cocaine overdose deaths by the premorbid occurrence of hyperthermia. Body temperature has a high correlation to a disordered CNS, leading to
the loss of thermal regulation. DA receptors are known to play a role in regulating core body
temperature. Since hyperthermia is a clinical feature of cocaine delirium, Kosten and Kleber[16] have
speculated that death occurred due to a malfunction in DAergic control of thermoregulation.
Hypothermia receptors are known to be downregulated by high levels of intrasynaptic DA. Direct
application of intracerebral DA at first lowers body temperature; however, a subsequent "rebound"
in body temperature occurs about 1 h after discontinuing this stimulation.[39,40]

When cocaine is repeatedly administered, DAergic receptor numbers are altered.[40,41] The likelihood of hyperthermia may be increased with chronic cocaine abuse if the DAergic receptors
involved in thermoregulation are undergoing adaptive changes with chronic cocaine exposure. In
keeping with this hypothesis, cocaine delirium victims had a different profile of D_2 receptor binding

within the thermoregulatory centers of the hypothalamus as compared to cocaine overdose deaths.[25,42] The density of the D_2 DA receptor subtype in the anterior and preoptic nuclei of the hypothalamus in the cocaine delirium subgroup of cocaine overdose deaths was decreased significantly ($p < 0.05$). These results may be relevant to an understanding of the contribution of selective alterations in D_1 and D_2 receptor subtypes in central DAergic temperature regulation. D_1 and D_2 receptors mediate opposite effects on thermoregulation, with the D_1 receptor mediating a prevailing increase in core body temperature, while the D_2 receptor mediates an opposing decrease in temperature.[42] Thus, the selective downregulation in the density of the D_2 DAergic receptor subtype within the hypothalamus may explain the loss of temperature regulation in cocaine delirium victims.

6.4 CONCLUSIONS

Cocaine abuse is associated with neuropsychiatric disorders, including acute psychotic episodes, paranoid states, and delirium. The mechanistic basis of these brain states is not fully known. The advent of new tools from the neurosciences and molecular genetics has led to a proliferation of research approaches aimed at defining the neurobiological consequences of chronic cocaine use. The development of radioligands with high specific activity and selectivity for neurotransmitter carriers and receptor subtypes has made it possible to map and quantify the neurochemical pathology in the brains of cocaine abusers. Since the DA transport carrier is a key regulator of DAergic neurotransmission, alterations in the numbers of these reuptake sites by cocaine may affect the balance in DAergic signaling. Understanding the influence of cocaine's effects on DAergic neurotransmission may shed light on the etiology of neuropsychiatric syndromes associated with cocaine abuse and dependence.

ACKNOWLEDGMENTS

The authors acknowledge the expert technical assistance of Margaret Basile, M.S., and Qinjie Ouyang, B.A. This work was supported by USPHS Grant DA06627.

REFERENCES

1. Baker, S.P. *The Injury Fact Book*. 2nd ed. Oxford University Press, New York, 1992.
2. Gawin, F.H. Cocaine addiction: psychology and neurophysiology. *Science*. 251: 1580, 1991.
3. Chitwood, D. Patterns and consequences of cocaine use. *Natl. Inst. Drug Abuse Res. Monogr. Ser.* 61: 111, 1985.
4. Gawin, F.H. and Kleber, H.D. Abstinence symptomatology and psychiatric diagnosis in cocaine abusers. *Arch. Gen. Psychiatry.* 43: 107, 1986.
5. Crawford, V. Comorbidity of substance misuse and psychiatric disorders. *Curr. Opin. Psychiatry.* 9: 231, 1996.
6. Kilbey, M.M., Breslau, N., and Andreski, P. Cocaine use and dependence in young adults: associated psychiatric disorders and personality traits. *Drug Alcohol Depend.* 29: 283, 1992.
7. Reiger, D.A., Farmer, M.E., Rae, D.S., Locke, B.Z., Keith, S.J., Judd, L.L., and Goodwin, F.K. Comorbidity of mental disorders with alcohol and other drug abuse. *J. Am. Med. Assoc.* 264: 2511, 1990.
8. Poole, R. and Brabbins, C. Drug induced psychosis. *Br. J. Psychiatry.* 168: 135, 1996.
9. Satel, S.L., Seibyl, J.P., and Charney, D.S. Prolonged cocaine psychosis implies underlying major psychopathology. *J. Clin. Psychiatry.* 52: 8, 1991.
10. Satel, S.L., Southwick, S.M., and Gawin, F.H. Clinical Features of cocaine-induced paranoia. *Am. J. Psychiatry.* 148: 495, 1991.

11. Angrist, B.M. Cocaine in the context of prior central nervous system stimulant epidemics. In *Cocaine in the Brain* (Mind in Medicine Series), Volkow, N. and Swann, A.C., Eds. Rutgers University Press, New Brunswick, NJ, 1990, 7.

12. Tucker, G.J. Delirium. *Semin. Clin. Neuropsychiatry.* 5: 63–255, 2000.

13. Wetli, C.V. and Fishbain, D.A. Cocaine-induced psychosis and sudden death in recreational cocaine users. *J. Forensic. Sci.* 30: 873, 1985.

14. Stephens, B.G., Baselt, R., Jentzen, J.M., Karch, S., Mash, D.C., and Wetli, C.V. Criteria for the interpretation of cocaine levels in human biological samples and their relation to cause of death. *J. Forensic Med. Pathol.* 25: 1, 2004.

15. Stephens, B.G., Jentzen, J.M., Karch, S., Wetli, C.V., and Mash, D.C. National Association of Medical Examiners position paper on the certification of cocaine-related deaths. *J. Forensic Med. Pathol.* 25: 11, 2004.

16. Kosten, T. and Kleber, H.D. Sudden death in cocaine abusers: relation to neuroleptic malignant syndrome. *Lancet.* 1: 1198, 1987.

17. Friedman, J.H., Feinberg, S.S., and Feldman, R.G. A neuroleptic malignant like syndrome due to levodopa therapy withdrawal. *J. Am. Med. Assoc.* 254: 2792, 1985.

18. Levison, J. Neuroleptic malignant syndrome. *Am. J. Psychiatry.* 142: 1137, 1985.

19. Kosten, T.R. and Kleber, H.D. Rapid death during cocaine abuse: a variant of neuroleptic malignant syndrome. *Am J. Drug Alcohol Abuse.* 14: 335, 1988.

20. Ruttenber, A.J., Lawler-Haevener, J., Wetli, C.V., Hearn, W.L., and Mash, D.C. Fatal excited delirium following cocaine use: epidemiologic findings provide evidence for new mechanisms of cocaine toxicity. *J. Forensic Toxicol.* 42: 25, 1997.

21. O'Halloran, R.L. and Lewman, L.V. Restraint asphyxiation in excited delirium. *Am J. Forensic Med. Pathol.* 14: 289, 1993.

22. Self, D.W. and Nestler, E.J. Relapse to drug-seeking: neural and molecular mechanisms. *Drug Alcohol Depend.* 51: 49, 1998.

23. Giros, B. and Caron, M.G. Molecular characterization of the dopamine transporter. *Trends. Pharmacol. Sci.* 14: 43, 1993.

24. Staley, J.K., Hearn, W.L., Ruttenber, A.J., Wetli, C.V., and Mash, D.C. High affinity cocaine recognition sites on the dopamine transporter are elevated in fatal cocaine overdose victims. *J. Pharmacol. Exp. Ther.* 271: 1678, 1995.

25. Staley, J.K., Wetli, C.V., Ruttenber, A.J., Hearn, W.L., Kung, H.F., and Mash, D.C. Dopamine transporter and receptor autoradiography in cocaine psychosis and sudden death. *Biol. Psychiatry.* 37: 656, 1995.

26. Little, K.Y., Zhang, L., McLaughlin, D.P., Desmond, T., Frey, K.A., Dalack, G.W., and Cassin, B.J. Striatal dopaminergic abnormalities in human cocaine users. *Am. J. Psychiatry.* 156: 238, 1999.

27. Mash, D.C., Pablo, J., Ouyang, Q., Hearn, W.L., and Izenwasser, S. Dopamine transport function is elevated in cocaine users. *J. Neurochem.* 81: 292, 2002.

28. Malison, R.T., Best, S.E., van Dyck, C.H., McCance, E.F., Wallace, E.A., Laruelle, M., Baldwin, R.M. Seibyl, J.P., Price, L.H., Kosten, T.R., and Innis, R.B. Elevated striatal dopamine transporters during acute cocaine abstinence as measured by [^{123}I]β-CIT SPECT. *Am. J. Psychiatry.* 155: 832, 1998.

29. Boja, J.W., Carroll, F.I., Rahman, M.A., Philip, A., Lewin, A.H., and Kuhar, M.J. New, potent cocaine analogs: ligand binding and transport studies in rat striatum. *Eur. J. Pharmacol.* 184: 329, 1990.

30. Staley, J.K., Boja, J.W., Carroll, F.I., Seltzman, H.H., Wyrick, C.D., Lewin, A.H., Abraham, P., and Mash, D.C. Mapping dopamine transporters in the human brain with novel selective cocaine analog [^{125}I]RTI-121. *Synapse.* 21: 364, 1995.

31. Madras, B.K., Spealman, R.D., Fahey, M.A., Neumeyer, J.L., Saha, J.K., and Milius, R.A. Cocaine receptors labeled by [^{3}H]2B-carbomethoxy-3β-(4-fluorophenyl)tropane. *Mol. Pharmacol.* 36: 518, 1989.

32. Mash, D.C. and Staley, J.K. Cocaine recognition sites on the human dopamine transporter in drug overdose victims. In *Neurotransmitter Transporter: Structure and Function*, Reith, M.E.A., Ed. Humana, New York, 1996, 56.

33. Alburges, M.E., Narang, N., and Wamsley, J.K. Alterations in the dopaminergic receptor system after chronic administration of cocaine. *Synapse.* 14: 314, 1993.

34. Wilson, J.M., Nobrega, J.N., Carroll, M.E., Niznik, H.B., Shannak, K., Lac, S.T., Pristupa, Z.B., Dixon, L.M., and Kish, S.J. Heterogenous subregional binding patterns of ^{3}H-WIN 35,428 and ^{3}H-GBR 12,935 are differentially regulated by chronic cocaine self-administration. *J. Neurosci.* 14: 2966, 1994.

35. Aloyo, V.J., Harvey, J.A., and Kirfides, A.L. Chronic cocaine increases WIN 35428 binding in rabbit caudate. *Soc. Neurosci. Abstr.* 19: 1843, 1994.

36. Letchworth, S.R., Nader, M., Smith, H., Friedman, D.P., and Porrino, L.J. Progression of changes in dopamine transporter binding site density as a result of cocaine self-administration in rhesus monkeys. *J. Neurosci.* 21: 2799, 2001.

37. Pristupa, Z.B., Wilson, J.M., Hoffman, B.J., Kish, S.J., and Niznik, H.B. Pharmacological heterogeneity of the cloned and native human dopamine transporter: dissociation of [^{3}H]WIN 35,428 and [^{3}H]GBR 12935 binding. *Mol. Pharmacol.* 45, 125, 1994.

38. Gelernter, J., Kranzler, H.R., Satel, S.L., and Rao, P.A. Genetic association between dopamine transporter protein alleles and cocaine-induced paranoia. *Neuropsychopharmacology.* 11: 195, 1994.

39. Costentin, J., Duterte-Boucher, D., Panissaud, C., and Michael-Titus, A. Dopamine D_1 and D_2 receptors mediate opposite effects of apomorphine on the body temperature of reserpinized mice. *Neuropharmacology.* 29: 31, 1990.

40. Meller, E., Hizami, R., and Kreuter, L. Hypothermia in mice: D_2 dopamine receptor mediation and absence of spare receptors. *Pharmacol. Biochem. Behav.* 32: 141, 1989.

41. Kleven, M.S., Perry, B.D., Woolverton, W.L., and Seiden, L.S. Effects of repeated injections of cocaine on D_1 and D_2 dopamine receptors in rat brain. *Brain Res.* 532: 265, 1990.

42. Wetli, C.V., Mash, D.C., and Karch, S. Agitated delirium and the neuroleptic malignant syndrome. *Am. J. Emerg. Med.* 14: 425, 1996.

CHAPTER 7

Neurobiology of 3,4-Methylenedioxymethamphetamine (MDMA, or "Ecstasy")

Michael H. Baumann, Ph.D. and Richard B. Rothman, M.D., Ph.D.
Clinical Psychopharmacology Section, Intramural Research Program, National Institute on Drug Abuse, National Institutes of Health, Department of Health and Human Services, Baltimore, Maryland

CONTENTS

3,4-Methylenedioxymethamphetamine (MDMA, or "Ecstasy") is an illicit drug used by young adults who attend "rave" dance parties in the U.S., Europe, and elsewhere. The allure of MDMA is related to its unique psychoactive effects, which include amphetamine-like stimulant actions, coupled with feelings of increased emotional sensitivity and closeness to others.[1,2] Epidemiological data indicate that MDMA misuse among children and adolescents is widespread in the U.S.[3,4] In

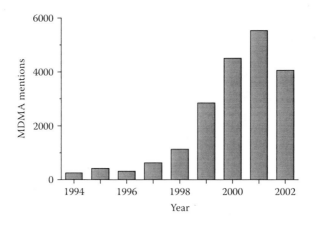

Figure 7.1   Emergency department mentions of MDMA from 1994–2002. (Adapted from Office of Applied Studies, SAMSHA, Drug Abuse Warning Network, 2002; updated 03/2003.)

a recent sampling of high school students, 10% of 12th graders reported using MDMA at least once.[5] As shown in Figure 7.1, MDMA-related emergency room visits have risen more than 20-fold in recent years, consistent with the increasing popularity of the drug. Serious adverse effects of acute MDMA intoxication include cardiac arrhythmias, hypertension, hyperthermia, serotonin (5-HT) syndrome, hyponatremia, liver problems, seizures, coma, and, in rare cases, death.[6] Accumulating evidence indicates that long-term MDMA abuse is associated with cognitive impairments and mood disturbances, which can last for months after cessation of drug intake.[7,8]

Despite the potential risks associated with illicit MDMA use, a growing number of clinicians believe the drug could have therapeutic potential in the treatment of psychiatric disorders.[9] For example, adjunct therapy with MDMA might prove useful for alleviating the anxiety that accompanies post-traumatic stress disorder (PTSD) or end-stage terminal illness. Indeed, clinical trials aimed at testing the efficacy of MDMA for the treatment of PTSD are under way.[10] MDMA has been administered to human subjects in controlled research settings, and few side effects are observed under these conditions, supporting the relative safety of the drug.[11,12] The aforementioned considerations provide compelling reasons to evaluate the neurobiology of MDMA and related compounds. In this chapter, we review the acute and long-term effects of MDMA administration on central nervous system (CNS) function. The chapter focuses on results obtained from rats since most preclinical MDMA research has been carried out in this animal model. Experimental data from our laboratory at NIDA are included to supplement literature reports, and clinical data are mentioned in certain instances to note similarities or differences between rats and humans.

7.1 MDMA INTERACTS WITH MONOAMINE TRANSPORTERS

7.1.1 *In Vitro* Studies

Figure 7.2 shows that MDMA is a ring-substituted analogue of methamphetamine, and "Ecstasy" tablets ingested by humans contain a racemic mixture of (+) and (−) isomers of the drug.[13,14] Upon systemic administration, MDMA is *N*-demethylated via first-pass metabolism in the liver to yield (+) and (−) isomers of the amphetamine analogue, 3,4-methylenedioxyamphetamine (MDA).[15,16] Initial pharmacological studies carried out in the 1980s revealed that isomers of MDMA and MDA stimulate efflux of 5-HT, and to a lesser extent dopamine (DA), in brain tissue preparations.[17–19] Subsequent investigations demonstrated that MDMA is a substrate for

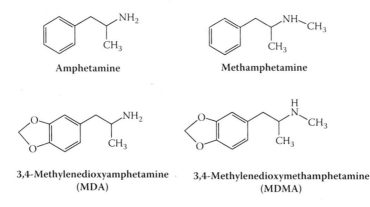

Figure 7.2   Chemical structures of MDMA and related compounds.

monoamine transporter proteins, evoking the non-exocytotic release of 5-HT, DA, and norepinephrine (NE) from nerve terminals.[20–22]

Like other substrate-type releasers, MDMA and MDA bind to plasma membrane monoamine transporters and are subsequently translocated into the cytoplasm.[23] The ensuing transmitter release occurs by a two-pronged mechanism: (1) transmitter molecules exit the cell along their concentration gradient via a diffusion-exchange process that involves reversal of normal transporter flux, and (2) cytoplasmic concentrations of transmitter are increased due to drug-induced disruption of vesicular storage.[24,25] This latter action serves to markedly increase the pool of cytoplasmic transmitter available for diffusion-exchange release. Because substrate-type releasing drugs must be transported into cells to promote transmitter release, transporter uptake inhibitors can block the effects of releasers. Figure 7.3 depicts data from our laboratory showing MDMA produces a dose-dependent release of preloaded [³H]5-HT and [³H]DA from rat brain synaptosomes. In these experiments, the "release" of preloaded radiolabeled transmitter is expressed as a reduction in the amount of tritium retained in tissue. Reserpine is added to the incubation medium to prevent the trapping of radiolabeled transmitter in vesicles.[26,27] MDMA-induced release of [³H]5-HT is antagonized by co-incubation with the selective 5-HT uptake inhibitor fluoxetine, whereas release of [³H]DA is antagonized by the selective DA uptake inhibitor GRB12909. These findings support

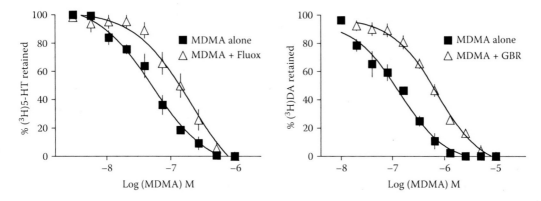

Figure 7.3   Dose–response effects of MDMA on the release of preloaded [³H]5-HT (left panel) and [³H]DA (right panel) from synaptosomes *in vitro*. [³H]Transmitter release is expressed as percent of tritium retained in tissue. Various concentrations of MDMA were incubated with or without the 5-HT uptake blocker fluoxetine (10 n*M*) in [³H]5-HT assays, whereas various concentrations of MDMA were incubated with or without the DA uptake blocker GBR12909 (10 n*M*) in [³H]DA assays. Data are mean ± SD for three separate experiments, each performed in triplicate. See Baumann et al.[39] for methods.

Table 7.1   Profile of MDMA and Related Compounds as Monoamine Transporter Substrates

Drug	5-HT Release EC_{50} (nM ± SD)	NE Release EC_{50} (nM ± SD)	DA Release EC_{50} (nM ± SD)
(+)-Methamphetamine	736 ± 45	12 ± 0.7	24 ± 2
(−)-Methamphetamine	4640 ± 240	29 ± 3	416 ± 20
(±)-MDMA	74.3 ± 5.6	136 ± 17	278 ± 12
(+)-MDMA	70.8 ± 5.2	110 ± 16	142 ± 6
(−)-MDMA	337 ± 34	564 ± 60	3682 ± 178
(+)-Amphetamine	1765 ± 94	7.1 ± 1.0	25 ± 4
(±)-MDA	159 ± 12	108 ± 12	290 ± 10
(+)-MDA	99.6 ± 7.4	98.5 ± 6.1	50.0 ± 8.0
(−)-MDA	313 ± 21	287 ± 23	900 ± 49

Sources: The data are taken from Partilla et al.[28] and Setola et al.[29] Details concerning *in vitro* methods can be found in these papers.

the hypothesis that MDMA stimulates 5-HT and DA release *in vitro* via interactions at 5-HT transporters (SERT) and DA transporters (DAT), respectively.

The data in Table 7.1 summarize structure–activity relationships for MDMA, MDA, and related drugs, with respect to monoamine release from rat brain synaptosomes.[28,29] Stereoisomers of MDMA and MDA are substrates for SERT, DAT, and NE transporters (NET), with (+) isomers exhibiting greater potency as releasers. In particular, (+) isomers of MDMA and MDA are much more effective DA releasers than their corresponding (−) isomers. It is noteworthy that (+) isomers of MDMA and MDA are rather nonselective in their ability to stimulate monoamine release *in vitro*. When compared to other amphetamines, the major effect of methylenedioxy ring-substitution is enhanced potency for 5-HT release and reduced potency for DA release. For example, (+)-MDMA releases 5-HT (EC_{50} = 70.8 nM) about ten times more potently than (+)-methamphetamine (EC_{50} = 736 nM), whereas (+)-MDMA releases DA (EC_{50} = 142 nM) about six times less potently than (+)-methamphetamine (EC_{50} = 24 nM).

7.1.2   *In Vivo* Microdialysis Studies

The technique of *in vivo* microdialysis allows continuous sampling of extracellular fluid from intact brain, and dialysate samples can be assayed for monoamine transmitters using various analytical methods.[30] Microdialysis studies in rats demonstrate that systemic administration of MDMA increases extracellular levels of 5-HT and DA in the brain, consistent with the *in vitro* results noted above.[31–34] Pretreatment with 5-HT uptake inhibitors can block the rise in dialysate 5-HT produced by MDMA, suggesting the involvement of SERT.[33,35] Interestingly, the effect of MDMA administration on dialysate DA appears to be more complex and entails at least two processes: (1) a tetrodotoxin-insensitive mechanism that involves substrate interaction at DAT proteins,[36,37] and (2) a tetrodotoxin-sensitive mechanism that involves activation of 5-HT$_{2A}$ receptor sites by endogenous 5-HT.[33,38] Findings from our laboratory, illustrated in Figure 7.4, reveal that intravenous (i.v.) MDMA administration causes dose-related elevations in extracellular levels of 5-HT and DA in rat nucleus accumbens.[39] In these experiments, drugs were administered to conscious rats undergoing *in vivo* microdialysis. Dialysate levels of 5-HT and DA were determined by high-performance liquid chromatography coupled to electrochemical detection (HPLC-ECD). We found that MDMA has greater effects on *in vivo* 5-HT release when compared to DA release, and this observation has been confirmed in brain regions such as cortex and striatum. At the 1 mg/kg i.v. dose of MDMA, extracellular 5-HT was elevated approximately sixfold above baseline whereas extracellular DA was elevated approximately twofold.

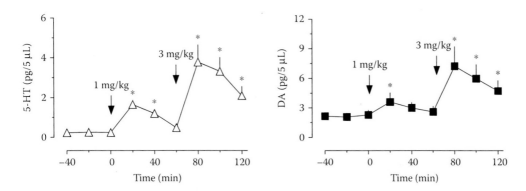

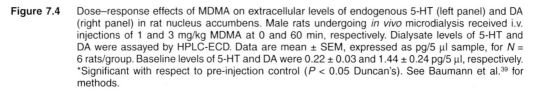

Figure 7.4   Dose–response effects of MDMA on extracellular levels of endogenous 5-HT (left panel) and DA (right panel) in rat nucleus accumbens. Male rats undergoing *in vivo* microdialysis received i.v. injections of 1 and 3 mg/kg MDMA at 0 and 60 min, respectively. Dialysate levels of 5-HT and DA were assayed by HPLC-ECD. Data are mean ± SEM, expressed as pg/5 μl sample, for N = 6 rats/group. Baseline levels of 5-HT and DA were 0.22 ± 0.03 and 1.44 ± 0.24 pg/5 μl, respectively. *Significant with respect to pre-injection control (P < 0.05 Duncan's). See Baumann et al.[39] for methods.

7.2 ACUTE EFFECTS OF MDMA

7.2.1   *In Vivo* Pharmacological Effects of MDMA

The acute CNS effects of MDMA administration are mediated by the release of monoamine transmitters, with the subsequent activation of presynaptic and postsynaptic receptor sites.[40] As specific examples in rats, MDMA suppresses 5-HT cell firing, evokes neuroendocrine secretion, and stimulates locomotor activity. MDMA-induced suppression of 5-HT cell firing in the dorsal and median raphe involves activation of presynaptic 5-HT$_{1A}$ autoreceptors by endogenous 5-HT.[41,42] Neuroendocrine effects of MDMA include secretion of prolactin from the anterior pituitary and corticosterone from the adrenal glands.[43] Evidence supports the notion that these MDMA-induced hormonal effects are mediated via postsynaptic 5-HT$_2$ receptors in the hypothalamus, which are activated by released 5-HT. MDMA elicits a unique profile of locomotor effects characterized by forward locomotion and elements of the 5-HT behavioral syndrome such as flattened body posture, Straub tail, and forepaw treading.[44-46] The complex motor effects of MDMA are dependent on monoamine release followed by activation of multiple postsynaptic 5-HT and DA receptor subtypes in the brain,[47] but the precise role of specific receptor subtypes is still under investigation.

In our laboratory, we carried out *in vivo* microdialysis in rats that were housed in chambers equipped with photo-beams to allow automated assessment of motor activity. Under these conditions, i.v. MDMA administration increases motor activity in conjunction with elevations in extracellular 5-HT and DA (see Figure 7.4). The data in Figure 7.5 demonstrate that MDMA increases forward locomotion (i.e., ambulation) and repetitive movements (i.e., stereotypy) in a dose-dependent fashion. Stereotypy produced by i.v. MDMA consists predominately of lateral side-to-side head weaving and reciprocal forepaw treading. We discovered that MDMA-induced 5-HT release in the nucleus accumbens and caudate nucleus is significantly correlated with stereotypic movements, whereas DA release in these brain regions is correlated with ambulation. These data suggest that 5-HT and DA systems influence MDMA-induced motor activation in a region-specific and modality-specific manner.

Adverse effects of high-dose MDMA intoxication, including cardiovascular stimulation and elevated body temperature, are thought to involve monoamine release from sympathetic nerves in the periphery or nerve terminals in the CNS.[48] MDMA increases heart rate and mean arterial

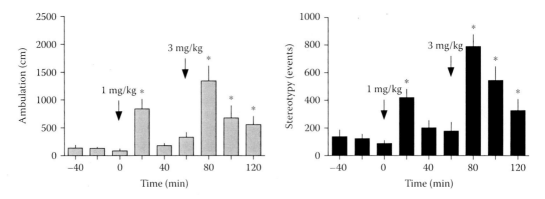

Figure 7.5   Dose–response effects of MDMA on ambulation (left panel) and stereotypy (right panel) in rats undergoing *in vivo* microdialysis. Male rats received i.v. injections of 1 and 3 mg/kg MDMA at 0 and 60 min, respectively. Ambulation (i.e., forward locomotion) and stereotypy (i.e., repetitive movements) were measured by photo-beam break analysis. Data are mean ± SEM, expressed as centimeters traveled for ambulation and number of events for stereotypy, with $N = 6$ rats/ group. *Significant with respect to pre-injection control ($P < 0.05$ Duncan's). See Baumann et al.[39] for methods.

pressure in conscious rats;[49,50] this cardiovascular stimulation is probably related to MDMA-induced release of peripheral NE stores, similar to the effects of amphetamine.[51] MDMA is reported to have weak agonist actions (i.e., $IC_{50} > 1$ μM) at α_2-adrenoreceptors and 5-HT$_2$ receptors, which might influence its cardiac and pressor effects.[52–54] Moreover, the MDMA metabolite MDA is a potent 5-HT$_{2B}$ agonist, and this property could contribute to adverse cardiovascular effects.[29] The ability of MDMA to elevate body temperature is well characterized in rats,[35,43,46,55] and this response has long been considered a 5-HT-mediated process. However, a recent study by Mechan et al.[35] provides convincing evidence that MDMA-induced hyperthermia in rats involves activation of postsynaptic D$_1$ receptors by released DA.

7.2.2   MDMA Metabolism

MDMA is extensively metabolized in humans and other species.[56] Figure 7.6 depicts the major pathway of MDMA biotransformation in humans, which entails: (1) *O*-demethylation catalyzed by cytochrome P450 2D6 (CYP2D6) and (2) *O*-methylation catalyzed by catechol-*O*-methyltransferase (COMT). CYP2D6 and COMT are both polymorphic in humans; the differential expression of CYP2D6 isoforms leads to marked inter-individual variations in the metabolism of serotonergic medications (e.g., SSRIs).[57] Interestingly, CYP2D6 is not present in rats, and this species expresses a homologous but functionally distinct cytochrome P450 2D1 that metabolizes MDMA.[58,59] A minor pathway of MDMA biotransformation in humans involves *N*-demethylation of MDMA to form MDA, which is subsequently *O*-demethylenated and *O*-methylated as described above. The *N*-demethylation pathway represents a more important mechanism for biotransformation of MDMA in rats when compared to humans.[60]

As noted above, MDA is a potent stimulator of monoamine release (see Table 7.1), and recent reports indicate that a number of MDMA metabolites are bioactive. For example, Forsling et al.[61] showed that the metabolite 4-hydroxy-3-methoxymethamphetamine (HMMA) is more potent than MDMA as a stimulator of vasopressin secretion from rat posterior pituitaries *in vitro*. The neuroendocrine effects produced by *in vivo* administration of MDMA metabolites have not been examined. Monks et al.[62] demonstrated that catechol metabolites of MDMA and MDA, namely, 3,4-dihydroxymethamphetamine (HHMA) and 3,4-dihydroxyamphetamine (HHA), exhibit neurotoxic properties when oxidized and conjugated with glutathione. Further characterization of the biological effects of MDMA metabolites is an important area of research.

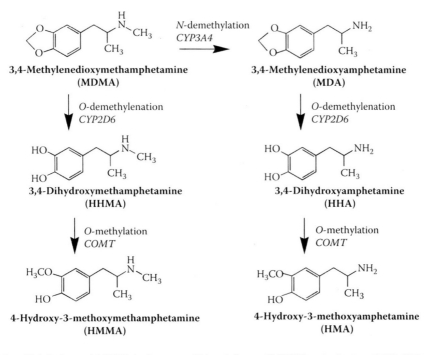

Figure 7.6 Metabolism of MDMA in humans. Abbreviations: *CYP2D6*, cytochrome P450 2D6; *CYP3A4*, cytochrome P450 3A4; *COMT*, catechol-*O*-methyltransferase. (Adapted from de la Torre and co-workers.[56])

The findings of de la Torre et al.[63] have shown that MDMA displays nonlinear kinetics in humans such that administration of increasing doses, or multiple doses, leads to unexpectedly high plasma levels of the drug. Enhanced plasma and tissue levels of MDMA are most likely related to auto-inhibition of MDMA metabolism, mediated via formation of a metabolite-enzyme complex that irreversibly inactivates CYP2D6.[64] Because of the nonlinear kinetics, repeated MDMA dosing could produce serious adverse consequences due to unusually high blood and tissue levels of the drug. The existing database of MDMA pharmacokinetic studies represents a curious situation where clinical findings are well documented, whereas preclinical data even in rodents are lacking. Specifically, few studies in animals have assessed the relationship between pharmacodynamic and pharmacokinetic effects of MDMA after single or repeated doses (although see Reference 65). No studies have systematically characterized the nonlinear kinetics of MDMA in rodent or nonhuman primate models.

7.3 LONG-TERM EFFECTS OF MDMA

7.3.1 Long-Term Effects of MDMA on 5-HT Neurons

The adverse effects of MDMA on 5-HT systems have been widely publicized, as many studies in animals show that high-dose MDMA administration produces persistent reductions in markers of 5-HT nerve terminal integrity.[66] Table 7.2 summarizes the findings of investigators who first demonstrated that MDMA causes long-term (>2 weeks) inhibition of tryptophan hydroxylase activity, depletion of brain tissue 5-HT, and reduction in SERT binding and function.[67–70] Immunohistochemical analysis of 5-HT in the CNS reveals an apparent loss of 5-HT axons and terminals in MDMA-treated rats, especially the fine-diameter projections arising from the dorsal raphe nucleus.[71,72] Moreover, the 5-HT axons and terminals remaining after MDMA treatment appear

Table 7.2   Long-Term Effects of MDMA on 5-HT Neuronal Markers in Rats

5-HT Deficit	Dose	Survival Interval	Ref.
Depletions of 5-HT in forebrain regions as measured by HPLC-ECD	10–40 mg/kg, s.c., twice daily, 4 days	2 weeks	Commins et al.[68]
Reductions in tryptophan hydroxylase activity in forebrain regions	10 mg/kg, s.c., single dose	2 weeks	Stone et al.[70]
Loss of [^{3}H]-paroxetine-labeled SERT binding sites in forebrain regions	20 mg/kg, s.c., twice daily, 4 days	2 weeks	Battaglia et al.[67]
Deceased immunoreactive 5-HT in fine axons and nerve terminals	20 mg/kg, s.c., twice daily, 4 days	2 weeks	O'Hearn et al.[72]

swollen and fragmented, suggesting structural damage. Time-course studies indicate that MDMA-induced 5-HT depletion occurs in a biphasic manner, with a rapidly occurring acute phase followed by a delayed long-term phase.[69,70] In the acute phase, which lasts for the first few hours after drug administration, massive depletion of brain tissue 5-HT is accompanied by inactivation of tryptophan hydroxylase. By 24 h later, tissue 5-HT recovers to normal levels but tryptophan hydroyxylase activity remains diminished. In the long-term phase, which begins within 1 week and lasts for months, depletion of 5-HT is accompanied by sustained inactivation of tryptophan hydroxylase and loss of SERT binding and function.[73,74]

The findings in Table 7.2 have been replicated by many investigators, and the spectrum of decrements produced by MDMA administration is typically described as 5-HT "neurotoxicity." Possible mechanisms underlying MDMA-induced 5-HT deficits are not completely understood, but evidence suggests the involvement of free radicals, oxidative damage, and metabolic stress.[75–77] As noted above, there are increasing data to support a role for toxic MDMA metabolites in mediating the long-term serotonergic effects of the drug.[60,62] Most studies examining MDMA neurotoxicity in rats have employed intraperitoneal (i.p.) or subcutaneous (s.c.) injections of 10 mg/kg or higher, either as single or repeated treatments. Such MDMA dosing regimens are known to produce significant hyperthermia, which exacerbates 5-HT deficits.[78,79]

There are some caveats to the hypothesis that MDMA produces 5-HT neurotoxicity. O'Hearn et al.[71,72] showed that MDMA has no effect on 5-HT cell bodies in the dorsal raphe despite profound loss of 5-HT in forebrain projection areas. Accordingly, the effects of MDMA on 5-HT neurons are often referred to as "axotomy," to account for the fact that perikarya are not damaged. MDMA-induced reductions in 5-HT levels and SERT binding eventually recover,[73,74] suggesting that 5-HT terminals are not destroyed. Many drugs used clinically produce effects similar to MDMA. For instance, reserpine causes sustained depletions of brain tissue 5-HT; yet reserpine is not considered a neurotoxin.[80] Chronic administration of 5-HT selective reuptake inhibitors (SSRIs), like paroxetine and sertraline, leads to a marked loss of SERT binding and function analogous to MDMA, but these agents are important therapeutic drugs rather than neurotoxins.[81,82] In fact, Frazer and Benmansour[83] have suggested that sustained downregulation of SERT binding and function underlies the efficacy of SSRIs in the treatment of depression and other mood disorders. Finally, high-dose administration of SSRIs produces swollen, fragmented, and abnormal 5-HT terminals, which are indistinguishable from the effects of high-dose MDMA and other substituted amphetamines.[84]

The above-mentioned caveats raise a number of questions with regard to MDMA neurotoxicity. Of course, the most important question is whether MDMA abuse causes neurotoxic damage to 5-HT systems in humans. This complex issue is a matter of ongoing debate, which has been addressed by recent papers.[85–87] Clinical studies designed to critically evaluate the long-term effects of MDMA are hampered by a range of factors including comorbid psychopathology and polydrug abuse among MDMA users. Animal models afford the unique opportunity to evaluate the potential neurotoxic effects of MDMA administration without many of these complicating factors.

Table 7.3   Effects of MDMA on Established Markers of Neurotoxicity in Rats

CNS Marker	Dosing Regimen	Survival Interval	Ref.
No change in 5-HT cell firing in raphe nuclei	20 mg/kg, s.c., twice daily, 4 days	2 weeks	Gartside et al.[88]
Increased silver-positive staining in degenerating neurons	80 mg/kg, s.c., twice daily, 4 days 25–150 mg/kg, s.c., twice daily, 2 days	15–48 h 2 days	Commins et al.[68] Jensen et al.[92]
No reactive astrogliosis, as measured by a lack of change in levels of GFAP	10–30 mg/kg, s.c., twice daily, 7 days 20 mg/kg, s.c., twice daily, 4 days 7.5 mg/kg, i.p., 3 doses	2 days 3 days, 1 week 2 days, 2 weeks	*O'Callaghan et al.[96] *Pubill et al.[98] *Wang et al.[97,99]

* These investigators found no effect of MDMA on GFAP expression, at doses that significantly depleted 5-HT levels in brain tissue.

7.3.2   Long-Term Effects of MDMA on Markers of Neurotoxicity

It is well accepted that MDMA produces 5-HT depletions in rat CNS, but much less attention has been devoted to the effects of MDMA on established markers of neurotoxicity such as cell death, silver-positive staining, and reactive gliosis. Support for the hypothesis of MDMA-induced axotomy relies heavily on immunohistochemical analysis of 5-HT levels, which could produce misleading results if not validated by other methods. For example, MDMA-induced loss of 5-HT could be due to persistent adaptive changes in gene expression or protein function, reflecting a state of metabolic quiescence rather than neurotoxic damage. Table 7.3 summarizes the effects of MDMA on hallmark measures of neurotoxicity.

Anatomical evidence reveals that MDMA does not damage 5-HT cell bodies, and functional studies support this notion. 5-HT neurons in the dorsal raphe exhibit pacemaker-like firing, which can be recorded using electrophysiological techniques.[41,42] High-dose MDMA administration (20 mg/kg, s.c., twice daily, 4 days) has no lasting effects on 5-HT cell firing or action potential characteristics when recordings are carried out 2 weeks after drug pretreatment.[88] The electrophysiological data in MDMA-pretreated rats differ from the effects produced by the neurotoxin 5,7-dihydroxytryptamine (5,7-DHT). In 5,7-DHT-pretreated rats, 5-HT cell firing is dramatically decreased in the dorsal raphe, in conjunction with loss of 5-HT immunofluorescence.[89,90] Thus, 5,7-DHT produces reductions in 5-HT cell firing that are attributable to cell death, but MDMA does not.

Silver staining techniques are commonly used to identify neuronal degeneration,[91] and two studies have examined the ability of MDMA to affect silver-positive staining (i.e., argyrophilia) in rat CNS. Commins et al.[68] administered single or multiple s.c. doses of 80 mg/kg MDMA to male rats, whereas Jensen et al.[92] gave twice daily s.c. injections of 50 to 250 mg/kg. In both cases, MDMA-pretreated rats displayed dose-dependent increases in the number of silver-positive nerve terminals, axons, and cell bodies in various brain areas, with the most severe degeneration observed in frontoparietal cortex. These results provide direct strong support for MDMA-induced neurotoxicity, but certain factors must be considered when interpreting the data. First, massive daily doses of MDMA ranging from 80 to 500 mg/kg were utilized, and these doses far exceed those producing 5-HT depletions in rats (see Table 7.2). Second, both investigations noted the presence of argyrophilic cell bodies in the cortex of MDMA-treated rats. Because 5-HT cell bodies are not present in the cortex,[93] these damaged cells must be nonserotonergic. Finally, the pattern of MDMA-induced silver staining does not correspond to the pattern of 5-HT innervation or the pattern of 5-HT depletions. It seems that sufficiently high doses of MDMA can increase silver-positive staining but this does not reflect 5-HT neurotoxicity per se.

A universal response to cell damage in the CNS is hypertrophy of astrocytes.[94] This "reactive gliosis" is accompanied by enhanced expression of glial-specific structural proteins, like glial fibrillary acidic protein (GFAP). O'Callaghan et al.[95] verified that a wide range of neurotoxic chemicals increase the levels of GFAP in rat CNS, indicating this protein can be used as a sensitive marker of neuronal damage. These investigators administered twice daily s.c. injections of 10 to 30 mg/kg MDMA to rats for 7 consecutive days; under these conditions, MDMA produced large 5-HT depletions in forebrain without any changes in GFAP expression.[96] Effects of MDMA on GFAP expression have been compared to the effects of 5,7-DHT.[96,97] At doses of MDMA and 5,7-DHT that cause comparable 5-HT depletions, only 5,7-DHT increases GFAP. Several recent reports from our laboratory and others confirm that MDMA-induced 5-HT depletions are not associated with increased GFAP expression.[97–99] Taken together, the majority of data from rats indicate that doses of MDMA causing significant 5-HT depletions (i.e., single or repeated doses of 10 to 20 mg/kg) do not induce cell death, silver-positive staining, or glial activation, suggesting these doses may not cause neuronal damage.

7.4 INTERSPECIES SCALING AND MDMA DOSING

7.4.1 Allometric Scaling and MDMA Dosing Regimens

A major point of controversy relates to the relevance of MDMA doses administered to rats when compared to those self-administered by humans (see References 40 and 48). As noted above, MDMA regimens that produce 5-HT depletions in rats involve administration of one or more doses of 10 to 20 mg/kg, whereas the amount of "Ecstasy" abused by humans is one or two tablets of 80 to 100 mg, about 1 to 3 mg/kg. Based on principles of "interspecies scaling," some investigators have proposed that high noxious doses of MDMA in rats correspond to recreational doses in humans.[100] The concept of interspecies scaling is based on shared biochemical mechanisms among eukaryotic cells (e.g., aerobic respiration), and was initially developed to describe variations in basal metabolic rate (BMR) in animal species of different sizes.[101,102] In the 1930s, Kleiber derived what is now called the "allometric equation" to describe the relationship between BMR and body weight. The generic form of the allometric equation is $Y = aW^b$, where Y is the variable of interest, W is the body weight, a is the allometric coefficient, and b is the allometric exponent. In the case where Y is BMR, b is accepted to be 0.75. In agreement with predictions of the allometric equation, smaller animals are known to have faster metabolism, heart rates, and circulation times, leading to faster clearance of exogenously administered drugs.

Unfortunately, the allometric equation is not always a valid predictor of drug dosing across species, especially for those compounds that are extensively metabolized in the liver.[103,104] As outlined previously, MDMA is readily metabolized *in vivo* (see Figure 7.6).[56,59] There are significant species differences in the expression level and functional activity of cytochrome P450 isoforms involved in the metabolism of MDMA.[59,60] The potential for nonlinear kinetics complicates comparative aspects of MDMA metabolism, and no information is available concerning this phenomenon in diverse species. Additionally, brain tissue uptake of substituted amphetamines is much greater in rats than in humans,[105] suggesting rats could be more sensitive than humans to the effects of MDMA, rather than vice versa. Collectively, the available information indicates that allometric scaling can be used to extrapolate *physiological* variables across species, but this method cannot be used to predict idiosyncratic distribution and metabolism of exogenously administered MDMA in a given animal model.

7.4.2 Effect Scaling and MDMA Dosing Regimens

The limitations of allometric scaling led us to investigate the method of "effect scaling" as an alternative strategy for matching equivalent doses of MDMA in rats and humans. In this approach,

Table 7.4   Comparative Neurobiological Effects of MDMA Administration in Rats and Humans

CNS Effect	Dose in Rats	Dose in Humans
In vivo release of 5-HT and DA	2.5 mg/kg, i.p., Gudelsky et al.[33] 1 mg/kg, s.c., Kankaanpaa et al.[34]	*1.5 mg/kg p.o., Liechti et al.[1,106]
Secretion of prolactin and glucocorticoids	1–3 mg/kg, i.p., Nash et al.[43]	125 mg, p.o., Mas et al.[11] 1.5 mg/kg, p.o., Harris et al.[12]
Drug discrimination	1.5 mg/kg, i.p., Schechter[108] 1.5 mg/kg, i.p., Glennon and Higgs[107]	1.5 mg/kg, p.o., Johanson et al.[109]
Drug self-administration	1 mg/kg, i.v., Schenck et al.[110]	**1–2 mg/kg, p.o., Tancer and Johanson[111]

* Subjective effects were attenuated by 5-HT uptake blockers, suggesting the involvement of transporter-mediated 5-HT release.
** Reinforcing effects were determined based on a multiple choice procedure.

the lowest dose of drug that produces specific pharmacological responses is determined for rats and humans, and subsequent dosing regimens in rats are calculated with reference to the predetermined threshold dose. Table 7.4 shows the doses of MDMA that produce comparable CNS effects in rats and humans. Remarkably, the findings reveal that doses of MDMA in the range of 1 to 2 mg/kg produce pharmacological effects that are equivalent in both species.

Administration of MDMA at doses of 1 to 3 mg/kg causes marked elevations in extracellular 5-HT and DA in rat brain, as determined by *in vivo* microdialysis.[33,34,39] Although it is impossible to directly measure 5-HT and DA release in living human brain, clinical studies indicate that subjective effects of MDMA (1.5 mg/kg, p.o.) are antagonized by SSRIs, suggesting the involvement of transporter-mediated release of 5-HT.[1,106] Nash et al.[43] showed that i.p. injections of 1 to 3 mg/kg of MDMA stimulate prolactin and corticosterone secretion in rats, and similar oral doses increase plasma prolactin and cortisol in human drug users.[11,12] The dose of MDMA discriminated by rats and humans is identical: 1.5 mg/kg, i.p., for rats[107,108] and 1.5 mg/kg, p.o., for humans.[109] Schenk et al.[110] demonstrated that rats can be trained to self-administer MDMA using i.v. doses ranging from 0.25 to 1.0 mg/kg, indicating these doses possess reinforcing efficacy. Tancer and Johanson[111] reported that 1 and 2 mg/kg doses of MDMA have reinforcing properties in humans that resemble those of (+)-amphetamine. The findings summarized in Table 7.4 indicate there is no need to use interspecies scaling to "adjust" MDMA doses between rats and humans.

Based on this analysis, we devised a repeated MDMA dosing regimen in rats to mimic a one-time recreational binge in humans. Male rats weighing 300 to 350 g served as subjects and were double-housed in plastic shoebox cages. In our initial studies, 3 i.p. injections of 1.5 or 7.5 mg/kg MDMA were administered, one dose every 2 h, to yield cumulative doses of 4.5 or 22.5 mg/kg, respectively. Control rats received saline vehicle according to the same schedule. Rats were removed from their cages to receive i.p. injections, but were otherwise confined to their home cages. The 1.5 mg/kg dose was used as a low "behavioral" dose whereas the 7.5 mg/kg dose was used as a high "noxious" dose (i.e., a dose fivefold greater than threshold). Our repeated dosing regimen was designed to account for the common practice of sequential dosing (i.e., "bumping") used by human subjects during rave parties. During the MDMA dosing procedure, rectal temperatures were recorded and 5-HT-mediated behaviors were scored every hour. Rats were decapitated 2 weeks after dosing, brain regions were dissected, and tissue levels of 5-HT and DA were determined by HPLC-ECD as described previously.[112]

Data in Figure 7.7 illustrate that repeated i.p. doses of 7.5 mg/kg MDMA elicit persistent hyperthermia on the day of treatment, whereas repeated doses of 1.5 mg/kg do not. As shown in Figure 7.8, high-dose MDMA treatment produces long-term depletions of tissue 5-HT in a number of brain regions (~50% reductions), but the low-dose group displays 5-HT concentrations similar to saline controls. Transmitter depletion is selective for 5-HT neurons since tissue DA levels are unaffected. The magnitude of 5-HT depletions depicted in Figure 7.8 is similar to that observed by others.[67-70] Our findings demonstrate that repeated injections of MDMA at a threshold behavioral

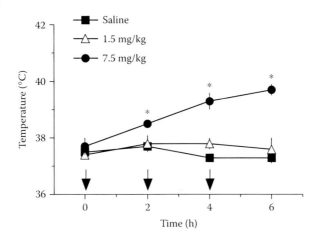

Figure 7.7   Acute effects of MDMA on core body temperature in rats. Male rats received three sequential i.p.
injections of 1.5 or 7.5 mg/kg MDMA, one dose every 2 h (i.e., injections at 0, 2, and 4 h). Saline
was administered on the same schedule. Core temperature was recorded via a rectal thermometer
probe every 2 h. Data are mean ± SEM expressed as degrees Celsius for $N = 5$ rats/group.
*Significant with respect to saline-injected control at each time point ($P < 0.05$ Duncan's).

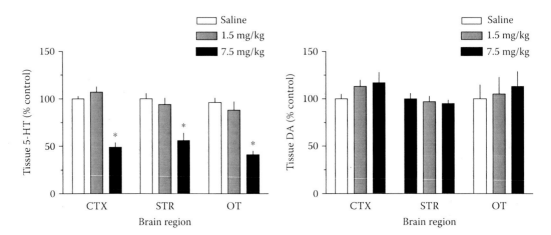

Figure 7.8   Long-term effects of MDMA on tissue levels of 5-HT (left panel) and DA (right panel) in brain
regions. Male rats received three i.p. injections of 1.5 or 7.5 mg/kg MDMA, one dose every 2 h.
Saline was administered on the same schedule. Rats were killed 2 weeks after injections, brain
regions were dissected, and tissue 5-HT and DA were assayed by HPLC-ECD.[112] Data are mean
± SEM expressed as percent of saline-treated control values for each region, $N = 5$ rats/group.
Control values of 5-HT and DA were 557 ± 24 and 28 ± 4 pg/mg tissue for frontal cortex (CTX),
429 ± 36 and 10,755 ± 780 pg/mg tissue for striatum (STR), and 1174 ± 114 and 4545 ± 426
pg/mg tissue for olfactory tubercle (OT). * Significant compared to saline-injected control for each
region ($P < 0.05$ Duncan's).

dose do not cause acute hyperthermia or long-term 5-HT depletions. In contrast, repeated injections
of MDMA at a dose that is fivefold higher than the behavioral dose induce both of these adverse
effects. The data are consistent with those of O'Shea et al.,[113] who reported that high-dose MDMA
(10 or 15 mg/kg, i.p.), but not low-dose MDMA (4 mg/kg, i.p.), causes acute hyperthermia and
long-term 5-HT depletion in Dark Agouti rats. Thus, our data confirm that acute hyperthermia
produced by MDMA is an important factor contributing to the mechanism underlying subsequent
long-term 5-HT depletion.

Table 7.5   Long-Term Effects of MDMA on Functional Indices of 5-HT Transmission in Rats

CNS Effect	Dosing Regimen	Survival Interval	Ref.
Reductions in evoked 5-HT release *in vivo*	20 mg/kg, s.c., twice daily, 4 days 10 mg/kg, i.p., twice daily, 4 days	2 weeks 1 week	Series et al.[114] Shankaran and Gudelsky[115]
Changes in corticosterone and prolactin secretion	20 mg/kg, s.c. 20 mg/kg, s.c., twice daily, 4 days	2 weeks 4, 8, and 12 months	Poland et al.[124,125] Poland et al.[125]
Impairments in short-term memory	10–20 mg/kg, s.c., twice daily, 3 days	2 weeks	*Marston et al.[134]
Increased anxiety-like behaviors	5 mg/kg, s.c., 1 or 4 doses, 2 days 7.5 mg/kg, s.c., twice daily, 3 days	3 months 2 weeks	**Morley et al.[135]; McGregor et al.[138] **Fone et al.[137]

* Most studies show no effect of MDMA on learning and memory in rats (see text).
** These investigators noted marked increases in anxiogenic behaviors in the absence of significant MDMA-induced 5-HT depletion in brain.

7.5 CONSEQUENCES OF MDMA-INDUCED 5-HT DEPLETIONS

As noted above, high-dose MDMA administration causes persistent inactivation of tryptophan hydroxylase, which leads to inhibition of 5-HT synthesis and long-term loss of 5-HT.[70,72] Moreover, MDMA-induced reduction in the density of SERT binding sites leads to decreased capacity for reuptake of [^{3}H]5-HT in nervous tissue.[67–69] Regardless of whether these deficits reflect neurotoxic damage or long-term adaptation, such changes would be expected to have discernible *in vivo* correlates. Many investigators have examined functional consequences of high-dose MDMA administration, and a comprehensive review of this subject is beyond the scope of the present review.[48] Nonetheless, the following discussion will consider long-term effects of MDMA (i.e., >2 weeks) on *in vivo* indicators of 5-HT function in rats, as measured by microdialysis sampling, neuroendocrine secretion, and specific aspects of behavior. A number of key findings are summarized in Table 7.5. In general, few published studies have been able to relate the magnitude of MDMA-induced 5-HT depletion to the degree of specific functional impairment. MDMA administration rarely causes persistent changes in baseline measures of neural function, and deficits are most readily demonstrated by provocation of the 5-HT system by pharmacological (e.g., drug challenge) or physiological means (e.g., environmental stress).

7.5.1 *In Vivo* Microdialysis Studies

In vivo microdialysis has been used to evaluate the persistent neurochemical consequences of MDMA exposure in rats.[88,114–116] Series et al.[114] carried out microdialysis in rat frontal cortex 2 weeks after a 4-day regimen of 20 mg/kg s.c. MDMA. Prior MDMA exposure did not affect baseline extracellular levels of 5-HT, but decreased levels of the 5-HT metabolite, 5-hydroxyindoleacetic acid (5-HIAA), to ~30% of control. Moreover, the ability of (+)-fenfluramine to evoke 5-HT release was markedly blunted in MDMA-pretreated rats. In an analogous investigation, Shankaran and Gudelsky[115] assessed neurochemical effects of acute MDMA challenge in rats that had previously received 4 doses of 10 mg/kg i.p. MDMA. A week after MDMA pretreatment, baseline levels of dialysate 5-HT and DA in striatum were not altered even though tissue levels of 5-HT were depleted by 50%. The ability of MDMA to evoke 5-HT release was severely impaired in MDMA-pretreated rats while the concurrent DA response was normal. In this same study, effects of MDMA on body temperature and 5-HT syndrome were attenuated in MDMA-pretreated rats, suggesting drug tolerance.

Taken together, the microdialysis data reveal several important consequences of MDMA administration: (1) baseline levels of dialysate 5-HT are unaltered, despite depletion of tissue indoles, (2) baseline levels of dialysate 5-HIAA are consistently decreased, and (3) stimulated release of 5-HT is blunted in response to pharmacological or physiological provocation. The microdialysis findings in MDMA-pretreated rats resemble those obtained with 5,7-DHT, in which drug-pretreated rats display normal baseline extracellular 5-HT but decreased 5-HIAA.[117–119] In a representative study, Kirby et al.[117] performed microdialysis in rat striatum 4 weeks after intracerebroventricular 5,7-DHT. These investigators found that reductions in baseline dialysate 5-HIAA and impairments in stimulated 5-HT release are highly correlated with the degree of tissue 5-HT depletion, whereas baseline dialysate 5-HT is not. In fact, depletions of brain tissue 5-HT up to 90% did not affect baseline levels of dialysate 5-HT. Clearly, adaptive mechanisms serve to maintain normal concentrations of synaptic 5-HT, even under conditions of severe transmitter depletion. A comparable situation exists after lesions of the nigrostriatal DA system in rats where baseline levels of extracellular DA are maintained in the physiological range despite substantial loss of tissue DA.[120] In the case of high-dose MDMA treatment, it seems feasible that reductions in 5-HT uptake (e.g., less functional SERT protein) and metabolism (e.g., decreased monoamine oxidase activity) can compensate for 5-HT depletions in order to keep optimal concentrations of 5-HT bathing nerve cells. On the other hand, deficits in the ability to release 5-HT are readily demonstrated in MDMA-pretreated rats when 5-HT systems are taxed by drug challenge or stressors.

7.5.2 Neuroendocrine Challenge Studies

5-HT neurons projecting to the hypothalamus provide stimulatory input for the secretion of adrenocorticotropin (ACTH) and prolactin from the anterior pituitary.[121] Accordingly, 5-HT releasers (e.g., fenfluramine) and 5-HT receptor agonists increase plasma levels of these hormones in rats and humans.[122] Neuroendocrine challenge experiments have identified changes in serotonergic responsiveness in rats treated with MDMA.[123–125] In the most comprehensive study, Poland et al.[125] examined effects of high-dose MDMA on hormone responses elicited by acute fenfluramine challenge. Rats received injections of 20 mg/kg s.c. MDMA and were tested 2 weeks later. Prior MDMA exposure did not alter baseline levels of circulating ACTH or prolactin. However, in MDMA-pretreated rats, fenfluramine-induced ACTH secretion was reduced while prolactin secretion was enhanced. The MDMA dosing regimen caused significant depletions of tissue 5-HT in various brain regions, including hypothalamus. In a follow-up time-course study, rats exposed to multiple doses of 20 mg/kg MDMA displayed blunted ACTH responses that persisted for 12 months, even though tissue levels of 5-HT were not depleted at this time point. The data show that high-dose MDMA can cause functional abnormalities for up to 1 year, and such changes are not necessarily coupled to 5-HT depletions.

In our laboratory, we wished to further explore the long-term neuroendocrine consequences of MDMA administration. Utilizing the "effect scaling" regimen described previously, male Sprague-Dawley rats received 3 i.p. injections of 1.5 or 7.5 mg/kg MDMA, one dose every 2 h. Control rats received saline vehicle according to the same schedule. A week after MDMA treatment, rats were fitted with indwelling jugular catheters under pentobarbital anesthesia. After 1 week of recovery from surgery (i.e., 2 weeks after MDMA or saline), rats were brought into the testing room, i.v. doses of 1 and 3 mg/kg MDMA were administered, and blood samples were withdrawn. Plasma levels of corticosterone and prolactin were measured by radioimmunoassay methods.[126] The data depicted in Figure 7.9 show that MDMA pretreatment did not alter baseline levels of either hormone. Acute administration of MDMA elicited dose-dependent elevations in circulating corticosterone and prolactin as shown by others.[43] Rats exposed to high-dose MDMA pretreatment displayed significant

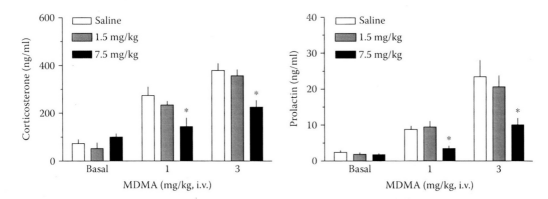

Figure 7.9   Effects of MDMA pretreatment on secretion of corticosterone (left panel) and prolactin (right panel) evoked by acute MDMA challenge. Male rats received three i.p. injections of 1.5 or 7.5 mg/kg MDMA, one dose every 2 h. Saline was administered on the same schedule. Then 2 weeks later rats received i.v. injections of 1 and 3 mg/kg MDMA. Blood samples were drawn via indwelling catheters; plasma corticosterone and prolactin were measured by RIA.[126] Data are mean ± SEM, expressed as ng/ml of plasma for N = 8 rats/group. Baseline corticosterone and prolactin levels were 73 ± 18 and 2.4 ± 0.6 ng/ml of plasma, respectively. *Significant compared to saline-pretreated control group ($P < 0.05$ Duncan's).

reductions in corticosterone and prolactin secretion in response to acute MDMA challenge, whereas hormone responses in the low-dose MDMA rats were indistinguishable from controls.

Our neuroendocrine results are consistent with the development of tolerance to hormonal effects of MDMA. These findings do not agree completely with the data of Poland et al.[125] discussed above. However, our findings are consistent with previous data showing blunted hormonal responses to fenfluramine in rats with fenfluramine-induced 5-HT depletions.[126] Perhaps more importantly, the data shown in Figure 7.9 are strikingly similar to clinical findings in which cortisol and prolactin responses to acute (+)-fenfluramine administration are reduced in human MDMA users.[85,127,128] Indeed, Gerra et al.[128] reported that (+)-fenfluramine-induced prolactin secretion is blunted in abstinent MDMA users for up to 1 year after cessation of drug use. The mechanism(s) underlying altered sensitivity to (+)-fenfluramine and MDMA are not known, but it is tempting to speculate that MDMA-induced impairments in evoked 5-HT release are involved, as shown by *in vivo* microdialysis studies. While some investigators have cited neuroendocrine changes in human MDMA users as evidence for 5-HT neurotoxicity, Gouzoulis-Mayfrank et al.[85] provide a compelling argument that endocrine abnormalities in MDMA users could be related to cannabis use rather than MDMA. Further experiments will be required to resolve the precise nature of neuroendocrine changes in MDMA users.

7.5.3   Behavioral Assessments

One of the more serious and disturbing clinical findings is that MDMA causes persistent cognitive deficits in human users.[7,8,87] Numerous studies have examined the effects of MDMA treatment on learning and memory in rats, and most studies failed to identify persistent impairments — even when extensive 5-HT depletions were present.[45,129–133] While an exhaustive review of this literature is not possible here, representative findings will be mentioned. In an extensive series of experiments, Seiden et al.[129] evaluated the effects of high-dose MDMA on a battery of tests including open-field behavior, schedule-controlled behavior, one-way avoidance, discriminated two-way avoidance, forced swim, and radial maze performance. Male rats received twice daily s.c. injections of 10 to 40 mg/kg MDMA for 4 days, and were tested beginning 2 weeks after treatment. Despite large depletions of brain tissue 5-HT, MDMA-pretreated rats exhibited normal behaviors in all paradigms. Likewise, Robinson et al.[130] found that MDMA-induced depletion of cortical 5-HT up to 70% did not alter spatial navigation, skilled forelimb use, or foraging behavior in rats. In contrast, Marston et al.[134] reported that MDMA administration produces persistent deficits in a delayed non-match to performance (DNMTP) procedure when long delay intervals are employed (i.e., 30 s). The authors theorized that delay-dependent impairments in the DNMTP procedure reflect MDMA-induced deficits in short-term memory consolidation, possibly attributable to 5-HT depletion.

With the exception of the findings of Marston et al., the collective behavioral data in rats indicate that MDMA-induced depletions of brain 5-HT have little or no effect on cognitive processes. There are several potential explanations for this apparent paradox. First, high-dose MDMA administration produces only partial depletion of 5-HT in the range of 40 to 60% in most brain areas. This level of 5-HT loss may not be sufficient to elicit behavioral alterations, as compensatory adaptations in 5-HT neurons could maintain normal physiological function. Second, MDMA appears to selectively affect fine diameter fibers arising from the dorsal raphe, and it seems possible that these 5-HT circuits may not subserve the behaviors being monitored. Third, the behavioral tests utilized in rat studies might not be sensitive enough to detect subtle changes in learning and memory processes. Finally, the functional reserve capacity in the CNS might be sufficient to compensate for even large depletions of a single transmitter.

While MDMA appears to have few long-term effects on cognition in rats, a growing body of evidence demonstrates that MDMA administration can cause persistent anxiety-like behaviors in this species.[135–137] Morley et al.[135] first reported that MDMA induces long-term anxiety in male rats. These investigators administered 1 or 4 i.p. injections of 5 mg/kg MDMA on 2 consecutive days, then tested rats 3 months later in a battery of anxiety-related paradigms including elevated plus maze, emergence, and social interaction tests. Rats receiving single or multiple MDMA injections displayed marked increases in anxiogenic behaviors in all three tests. In a follow-up study, Gurtman et al.[136] replicated the original findings of Morley et al. using rats pretreated with 4 i.p. injections of 5 mg/kg MDMA for 2 days — persistent anxiogenic effects of MDMA were associated with depletions of 5-HT in the amygdala, hippocampus, and striatum. Interestingly, Fone et al.[137] showed that administration of MDMA to adolescent rats caused anxiety-like impairments in social interaction, even in the absence of 5-HT depletions or reductions in [3H]-paroxetine-labeled SERT binding sites. These data suggest that MDMA-induced anxiety does not require 5-HT deficits.

In an attempt to determine potential mechanisms underlying MDMA-induced anxiety, McGregor et al.[138] evaluated effects of the drug on anxiety-related behaviors and a number of post-mortem parameters including autoradiography for SERT and 5-HT receptor subtypes. Rats received moderate (5 mg/kg, i.p., 2 days) or high (5 mg/kg, i.p., 4 injections, 2 days) doses of MDMA, and tests were conducted 10 weeks later. This study confirmed that moderate doses of MDMA can cause protracted increases in anxiety-like behaviors without significant 5-HT depletions. Furthermore, the autoradiographic analysis revealed that anxiogenic effects of MDMA may involve long-term reductions in 5-HT$_{2A/2C}$ receptors rather than reductions in SERT binding. Additional work by Bull et al.[139,140] suggests that decreases in the sensitivity of 5-HT$_{2A}$ receptors, but not 5-HT$_{2C}$ receptors, could underlie MDMA-associated anxiety. Clearly, more investigation into this important area of research is warranted.

7.6 CONCLUSIONS

The findings reviewed here allow a number of tentative conclusions to be made with regard to MDMA neurobiology. (1) MDMA is a substrate for monoamine transporters, and non-exocytotic release of 5-HT, NE, and DA underlies pharmacological effects of the drug. While MDMA is often considered a selective serotonergic agent, many actions including cardiovascular stimulation and hyperthermia likely involve NE and DA mechanisms. (2) MDMA produces long-term changes in 5-HT neurons, as exemplified by sustained depletions of forebrain 5-HT in rats. Emerging evidence indicates that 5-HT deficits are not synonymous with neuronal damage, however, since doses of MDMA that cause marked 5-HT depletions (e.g., 10 to 20 mg/kg) are not associated with cell death, silver-positive staining, or reactive gliosis. Like many other psychotropic agents, MDMA is capable of producing *bona fide* neurotoxicity at sufficient doses (e.g., >30 mg/kg) and damage is not confined to 5-HT neurons. (3) There appears to be no scientific rationale for using interspecies scaling to adjust doses of MDMA between rats and humans because behaviorally active doses are similar in both species (e.g., 1 to 2 mg/kg). Nonetheless, the complex metabolism of MDMA needs to be examined in various animal species to permit comparison with clinical literature and to validate appropriate preclinical models. (4) MDMA-induced 5-HT depletions in rats are accompanied by abnormalities in evoked 5-HT release, neuroendocrine secretion, and specific behaviors. The clinical relevance of preclinical findings is uncertain, but the fact that MDMA can produce persistent increases in anxiety-like behaviors in rats without measurable 5-HT deficits suggests even moderate doses may pose risks.

ACKNOWLEDGMENTS

This research was generously supported by the NIDA Intramural Research Program. The authors are indebted to John Partilla, Chris Dersch, Mario Ayestas, Robert Clark, Fred Franken, and John Rutter for their expert technical assistance during these studies.

REFERENCES

1. Liechti, M.E. and Vollenweider, F.X., Which neuroreceptors mediate the subjective effects of MDMA in humans? A summary of mechanistic studies, *Hum. Psychopharmacol.* 16(8), 589–598, 2001.
2. Vollenweider, F.X., Gamma, A., Liechti, M., and Huber, T., Psychological and cardiovascular effects and short-term sequelae of MDMA ("ecstasy") in MDMA-naive healthy volunteers, *Neuropsychopharmacology* 19(4), 241–251, 1998.
3. Landry, M.J., MDMA: a review of epidemiologic data, *J. Psychoactive Drugs* 34(2), 163–9, 2002.
4. Yacoubian, G.S., Jr., Tracking ecstasy trends in the United States with data from three national drug surveillance systems, *J. Drug Educ.* 33(3), 245–258, 2003.
5. Banken, J.A., Drug abuse trends among youth in the United States, *Ann. N.Y. Acad. Sci.* 1025, 465–471, 2004.
6. Schifano, F., A bitter pill. Overview of ecstasy (MDMA, MDA) related fatalities, *Psychopharmacology* (Berlin) 173(3–4), 242–248, 2004.
7. Morgan, M.J., Ecstasy (MDMA): a review of its possible persistent psychological effects, *Psychopharmacology* (Berlin) 152(3), 230–248, 2000.
8. Parrott, A.C., Recreational Ecstasy/MDMA, the serotonin syndrome, and serotonergic neurotoxicity, *Pharmacol. Biochem. Behav.* 71(4), 837–844, 2002.
9. Doblin, R., A clinical plan for MDMA (Ecstasy) in the treatment of posttraumatic stress disorder (PTSD): partnering with the FDA, *J. Psychoactive Drugs* 34(2), 185–94, 2002.
10. Check, E., Psychedelic drugs: the ups and downs of ecstasy, *Nature* 429(6988), 126–128, 2004.
11. Mas, M., Farre, M., de la Torre, R., Roset, P.N., Ortuno, J., Segura, J., and Cami, J., Cardiovascular and neuroendocrine effects and pharmacokinetics of 3,4-methylenedioxymethamphetamine in humans, *J. Pharmacol. Exp. Ther.* 290(1), 136–145, 1999.
12. Harris, D.S., Baggott, M., Mendelson, J.H., Mendelson, J.E., and Jones, R.T., Subjective and hormonal effects of 3,4-methylenedioxymethamphetamine (MDMA) in humans, *Psychopharmacology* (Berlin) 162(4), 396–405, 2002.
13. Bell, S.E., Burns, D.T., Dennis, A.C., and Speers, J.S., Rapid analysis of ecstasy and related phenethylamines in seized tablets by Raman spectroscopy, *Analyst* 125(3), 541–544, 2000.
14. Huang, Y.S., Liu, J.T., Lin, L.C., and Lin, C.H., Chiral separation of 3,4-methylenedioxymethamphetamine and related compounds in clandestine tablets and urine samples by capillary electrophoresis/fluorescence spectroscopy, *Electrophoresis* 24(6), 1097–1104, 2003.

15. Cho, A.K., Hiramatsu, M., Distefano, E.W., Chang, A.S., and Jenden, D.J., Stereochemical differences in the metabolism of 3,4-methylenedioxymethamphetamine *in vivo* and *in vitro*: a pharmacokinetic analysis, *Drug Metab. Dispos.* 18(5), 686–691, 1990.

16. Lim, H.K. and Foltz, R.L., *In vivo* and *in vitro* metabolism of 3,4-(methylenedioxy)methamphetamine in the rat: identification of metabolites using an ion trap detector, *Chem. Res. Toxicol.* 1(6), 370–378, 1988.

17. Nichols, D.E., Lloyd, D.H., Hoffman, A.J., Nichols, M.B., and Yim, G.K., Effects of certain hallucinogenic amphetamine analogues on the release of [^{3}H]serotonin from rat brain synaptosomes, *J. Med. Chem.* 25(5), 530–535, 1982.

18. Johnson, M.P., Hoffman, A.J., and Nichols, D.E., Effects of the enantiomers of MDA, MDMA and related analogues on [^{3}H]serotonin and [^{3}H]dopamine release from superfused rat brain slices, *Eur. J. Pharmacol.* 132(2–3), 269–276, 1986.

19. Schmidt, C.J., Levin, J.A., and Lovenberg, W., *In vitro* and *in vivo* neurochemical effects of methylenedioxymethamphetamine on striatal monoaminergic systems in the rat brain, *Biochem. Pharmacol.* 36(5), 747–755, 1987.

20. Fitzgerald, J.L. and Reid, J.J., Interactions of methylenedioxymethamphetamine with monoamine transmitter release mechanisms in rat brain slices, *Naunyn Schmiedeberg's Arch. Pharmacol.* 347(3), 313–323, 1993.

21. Berger, U.V., Gu, X.F., and Azmitia, E.C., The substituted amphetamines 3,4-methylenedioxymethamphetamine, methamphetamine, *p*-chloroamphetamine and fenfluramine induce 5-hydroxytryptamine release via a common mechanism blocked by fluoxetine and cocaine, *Eur. J. Pharmacol.* 215(2–3), 153–160, 1992.

22. Crespi, D., Mennini, T., and Gobbi, M., Carrier-dependent and Ca(2+)-dependent 5-HT and dopamine release induced by (+)-amphetamine, 3,4-methylendioxymethamphetamine, *p*-chloroamphetamine and (+)-fenfluramine, *Br. J. Pharmacol.* 121(8), 1735–1743, 1997.

23. Rothman, R.B. and Baumann, M.H., Therapeutic and adverse actions of serotonin transporter substrates, *Pharmacol. Ther.* 95(1), 73–88, 2002.

24. Rudnick, G. and Wall, S.C., The molecular mechanism of "ecstasy" [3,4-methylenedioxy-methamphetamine (MDMA)]: serotonin transporters are targets for MDMA-induced serotonin release, *Proc. Natl. Acad. Sci. U.S.A.* 89(5), 1817–1821, 1992.

25. Schuldiner, S., Steiner-Mordoch, S., Yelin, R., Wall, S.C., and Rudnick, G., Amphetamine derivatives interact with both plasma membrane and secretory vesicle biogenic amine transporters, *Mol. Pharmacol.* 44(6), 1227–1231, 1993.

26. Rothman, R.B., Baumann, M.H., Dersch, C.M., Romero, D.V., Rice, K.C., Carroll, F.I., and Partilla, J.S., Amphetamine-type central nervous system stimulants release norepinephrine more potently than they release dopamine and serotonin, *Synapse* 39(1), 32–41, 2001.

27. Rothman, R.B., Partilla, J.S., Baumann, M.H., Dersch, C.M., Carroll, F.I., and Rice, K.C., Neurochemical neutralization of methamphetamine with high-affinity nonselective inhibitors of biogenic amine transporters: a pharmacological strategy for treating stimulant abuse, *Synapse* 35(3), 222–227, 2000.

28. Partilla, J.S., Dersch, C.M., Yu, H., Rice, K.C., and Rothman, R.B., Neurochemical neutralization of amphetamine-type stimulants in rat brain by the indatraline analog (–)-HY038, *Brain Res. Bull.* 53(6), 821–826, 2000.

29. Setola, V., Hufeisen, S.J., Grande-Allen, K.J., Vesely, I., Glennon, R.A., Blough, B., Rothman, R.B., and Roth, B.L., 3,4-Methylenedioxymethamphetamine (MDMA, "Ecstasy") induces fenfluramine-like proliferative actions on human cardiac valvular interstitial cells *in vitro*, *Mol. Pharmacol.* 63(6), 1223–1229, 2003.

30. Baumann, M.H. and Rutter, J.J., Application of *in vivo* microdialysis methods to the study of psychomotor stimulant drugs, in *Methods in Drug Abuse Research, Cellular and Circuit Level Analysis*, Warerhouse, B.D., Ed., CRC Press, Boca Raton, FL, 2003, 51–86.

31. Nash, J.F. and Nichols, D.E., Microdialysis studies on 3,4-methylenedioxyamphetamine and structurally related analogues, *Eur. J. Pharmacol.* 200(1), 53–58, 1991.

32. Yamamoto, B.K., Nash, J.F., and Gudelsky, G.A., Modulation of methylenedioxymethamphetamine-induced striatal dopamine release by the interaction between serotonin and gamma-aminobutyric acid in the substantia nigra, *J. Pharmacol. Exp. Ther.* 273(3), 1063–1070, 1995.

33. Gudelsky, G.A. and Nash, J.F., Carrier-mediated release of serotonin by 3,4-methylenedioxymethamphetamine: implications for serotonin-dopamine interactions, *J. Neurochem.* 66(1), 243–249, 1996.

34. Kankaanpaa, A., Meririnne, E., Lillsunde, P., and Seppala, T., The acute effects of amphetamine derivatives on extracellular serotonin and dopamine levels in rat nucleus accumbens, *Pharmacol. Biochem. Behav.* 59(4), 1003–1009, 1998.

35. Mechan, A.O., Esteban, B., O'Shea, E., Elliott, J.M., Colado, M.I., and Green, A.R., The pharmacology of the acute hyperthermic response that follows administration of 3,4-methylenedioxymethamphetamine (MDMA, "ecstasy") to rats, *Br. J. Pharmacol.* 135(1), 170–180, 2002.

36. Nash, J.F. and Brodkin, J., Microdialysis studies on 3,4-methylenedioxymethamphetamine-induced dopamine release: effect of dopamine uptake inhibitors, *J. Pharmacol. Exp. Ther.* 259(2), 820–825, 1991.

37. Shankaran, M., Yamamoto, B.K., and Gudelsky, G.A., Mazindol attenuates the 3,4-methylenedioxymethamphetamine-induced formation of hydroxyl radicals and long-term depletion of serotonin in the striatum, *J. Neurochem.* 72(6), 2516–2522, 1999.

38. Schmidt, C.J., Sullivan, C.K., and Fadayel, G.M., Blockade of striatal 5-hydroxytryptamine2 receptors reduces the increase in extracellular concentrations of dopamine produced by the amphetamine analogue 3,4-methylenedioxymethamphetamine, *J. Neurochem.* 62(4), 1382–1389, 1994.

39. Baumann, M.H., Clark, R.D., Budzynski, A.G., Partilla, J.S., Blough, B.E., and Rothman, R.B., N-substituted piperazines abused by humans mimic the molecular mechanism of 3,4-methylenedioxymethamphetamine (MDMA, or "Ecstasy"), *Neuropsychopharmacology* 30(3), 550–560, 2005.

40. Cole, J.C. and Sumnall, H.R., The pre-clinical behavioural pharmacology of 3,4-methylenedioxymethamphetamine (MDMA), *Neurosci. Biobehav. Rev.* 27(3), 199–217, 2003.

41. Sprouse, J.S., Bradberry, C.W., Roth, R.H., and Aghajanian, G.K., MDMA (3,4-methylenedioxymethamphetamine) inhibits the firing of dorsal raphe neurons in brain slices via release of serotonin, *Eur. J. Pharmacol.* 167(3), 375–383, 1989.

42. Gartside, S.E., McQuade, R., and Sharp, T., Acute effects of 3,4-methylenedioxymethamphetamine (MDMA) on 5-HT cell firing and release: comparison between dorsal and median raphe 5-HT systems, *Neuropharmacology* 36(11–12), 1697–703, 1997.

43. Nash, J.F., Jr., Meltzer, H.Y., and Gudelsky, G.A., Elevation of serum prolactin and corticosterone concentrations in the rat after the administration of 3,4-methylenedioxymethamphetamine, *J. Pharmacol. Exp. Ther.* 245(3), 873–879, 1988.

44. Gold, L.H., Koob, G.F., and Geyer, M.A., Stimulant and hallucinogenic behavioral profiles of 3,4-methylenedioxymethamphetamine and N-ethyl-3,4-methylenedioxyamphetamine in rats, *J. Pharmacol. Exp. Ther.* 247(2), 547–555, 1988.

45. Slikker, W., Jr., Holson, R.R., Ali, S.F., Kolta, M.G., Paule, M.G., Scallet, A.C., McMillan, D.E., Bailey, J.R., Hong, J.S., and Scalzo, F.M., Behavioral and neurochemical effects of orally administered MDMA in the rodent and nonhuman primate, *Neurotoxicology* 10(3), 529–542, 1989.

46. Spanos, L.J. and Yamamoto, B.K., Acute and subchronic effects of methylenedioxymethamphetamine [(+/–)MDMA] on locomotion and serotonin syndrome behavior in the rat, *Pharmacol. Biochem. Behav.* 32(4), 835–840, 1989.

47. Bankson, M.G. and Cunningham, K.A., 3,4-Methylenedioxymethamphetamine (MDMA) as a unique model of serotonin receptor function and serotonin-dopamine interactions, *J. Pharmacol. Exp. Ther.* 297(3), 846–852, 2001.

48. Green, A.R., Mechan, A.O., Elliott, J.M., O'Shea, E., and Colado, M.I., The pharmacology and clinical pharmacology of 3,4-methylenedioxymethamphetamine (MDMA, "ecstasy"), *Pharmacol. Rev.* 55(3), 463–508, 2003.

49. O'Cain, P.A., Hletko, S.B., Ogden, B.A., and Varner, K.J., Cardiovascular and sympathetic responses and reflex changes elicited by MDMA, *Physiol. Behav.* 70(1–2), 141–148, 2000.

50. Badon, L.A., Hicks, A., Lord, K., Ogden, B.A., Meleg-Smith, S., and Varner, K.J., Changes in cardiovascular responsiveness and cardiotoxicity elicited during binge administration of Ecstasy, *J. Pharmacol. Exp. Ther.* 302(3), 898–907, 2002.

51. Fitzgerald, J.L. and Reid, J.J., Sympathomimetic actions of methylenedioxymethamphetamine in rat and rabbit isolated cardiovascular tissues, *J. Pharm. Pharmacol.* 46(10), 826–832, 1994.

52. Lyon, R.A., Glennon, R.A., and Titeler, M., 3,4-Methylenedioxymethamphetamine (MDMA): stereoselective interactions at brain 5-HT1 and 5-HT2 receptors, *Psychopharmacology* (Berlin) 88(4), 525–526, 1986.

53. Battaglia, G. and De Souza, E.B., Pharmacologic profile of amphetamine derivatives at various brain recognition sites: selective effects on serotonergic systems, *NIDA Res. Monogr.* 94, 240–258, 1989.

54. Lavelle, A., Honner, V., and Docherty, J.R., Investigation of the prejunctional alpha2-adrenoceptor mediated actions of MDMA in rat atrium and vas deferens, *Br. J. Pharmacol.* 128(5), 975–980, 1999.

55. Dafters, R.I., Hyperthermia following MDMA administration in rats: effects of ambient temperature, water consumption, and chronic dosing, *Physiol. Behav.* 58(5), 877–882, 1995.

56. de la Torre, R., Farre, M., Roset, P.N., Pizarro, N., Abanades, S., Segura, M., Segura, J., and Cami, J., Human pharmacology of MDMA: pharmacokinetics, metabolism, and disposition, *Ther. Drug Monit.* 26(2), 137–144, 2004.

57. Charlier, C., Broly, F., Lhermitte, M., Pinto, E., Ansseau, M., and Plomteux, G., Polymorphisms in the CYP 2D6 gene: association with plasma concentrations of fluoxetine and paroxetine, *Ther. Drug Monit.* 25(6), 738–742, 2003.

58. Malpass, A., White, J.M., Irvine, R.J., Somogyi, A.A., and Bochner, F., Acute toxicity of 3,4-methylenedioxymethamphetamine (MDMA) in Sprague-Dawley and Dark Agouti rats, *Pharmacol. Biochem. Behav.* 64(1), 29–34, 1999.

59. Maurer, H.H., Bickeboeller-Friedrich, J., Kraemer, T., and Peters, F.T., Toxicokinetics and analytical toxicology of amphetamine-derived designer drugs ("Ecstasy"), *Toxicol. Lett.* March 15, 112–113, 133–142, 2000.

60. de la Torre, R. and Farre, M., Neurotoxicity of MDMA (ecstasy): the limitations of scaling from animals to humans, *Trends Pharmacol. Sci.* 25(10), 505–508, 2004.

61. Forsling, M.L., Fallon, J.K., Shah, D., Tilbrook, G.S., Cowan, D.A., Kicman, A.T., and Hutt, A.J., The effect of 3,4-methylenedioxymethamphetamine (MDMA, "ecstasy") and its metabolites on neurohypophysial hormone release from the isolated rat hypothalamus, *Br. J. Pharmacol.* 135(3), 649–656, 2002.

62. Monks, T.J., Jones, D.C., Bai, F., and Lau, S.S., The role of metabolism in 3,4-(+)-methylenedioxyamphetamine and 3,4-(+)-methylenedioxymethamphetamine (ecstasy) toxicity, *Ther. Drug Monit.* 26(2), 132–136, 2004.

63. de la Torre, R., Farre, M., Ortuno, J., Mas, M., Brenneisen, R., Roset, P.N., Segura, J., and Cami, J., Non-linear pharmacokinetics of MDMA ("ecstasy") in humans, *Br. J. Clin. Pharmacol.* 49(2), 104–109, 2000.

64. Wu, D., Otton, S.V., Inaba, T., Kalow, W., and Sellers, E.M., Interactions of amphetamine analogs with human liver CYP2D6, *Biochem. Pharmacol.* 53(11), 1605–1612, 1997.

65. Chu, T., Kumagai, Y., DiStefano, E.W., and Cho, A.K., Disposition of methylenedioxymethamphetamine and three metabolites in the brains of different rat strains and their possible roles in acute serotonin depletion, *Biochem. Pharmacol.* 51(6), 789–796, 1996.

66. Lyles, J. and Cadet, J.L., Methylenedioxymethamphetamine (MDMA, Ecstasy) neurotoxicity: cellular and molecular mechanisms, *Brain Res. Brain Res. Rev.* 42(2), 155–168, 2003.

67. Battaglia, G., Yeh, S.Y., O'Hearn, E., Molliver, M.E., Kuhar, M.J., and De Souza, E.B., 3,4-Methylenedioxymethamphetamine and 3,4-methylenedioxyamphetamine destroy serotonin terminals in rat brain: quantification of neurodegeneration by measurement of [^{3}H]paroxetine-labeled serotonin uptake sites, *J. Pharmacol. Exp. Ther.* 242(3), 911–916, 1987.

68. Commins, D.L., Vosmer, G., Virus, R.M., Woolverton, W.L., Schuster, C.R., and Seiden, L.S., Biochemical and histological evidence that methylenedioxymethylamphetamine (MDMA) is toxic to neurons in the rat brain, *J. Pharmacol. Exp. Ther.* 241(1), 338–345, 1987.

69. Schmidt, C.J., Neurotoxicity of the psychedelic amphetamine, methylenedioxymethamphetamine, *J. Pharmacol. Exp. Ther.* 240(1), 1–7, 1987.

70. Stone, D.M., Merchant, K.M., Hanson, G.R., and Gibb, J.W., Immediate and long-term effects of 3,4-methylenedioxymethamphetamine on serotonin pathways in brain of rat, *Neuropharmacology* 26(12), 1677–1683, 1987.

71. Molliver, M.E., Berger, U.V., Mamounas, L.A., Molliver, D.C., O'Hearn, E., and Wilson, M.A., Neurotoxicity of MDMA and related compounds: anatomic studies, *Ann. N.Y. Acad. Sci.* 600, 649–661; discussion 661–664, 1990.

72. O'Hearn, E., Battaglia, G., De Souza, E.B., Kuhar, M.J., and Molliver, M.E., Methylenedioxyamphet-amine (MDA) and methylenedioxymethamphetamine (MDMA) cause selective ablation of seroton-ergic axon terminals in forebrain: immunocytochemical evidence for neurotoxicity, *J. Neurosci.* 8(8), 2788–2803, 1988.

73. Battaglia, G., Yeh, S.Y., and De Souza, E.B., MDMA-induced neurotoxicity: parameters of degen-eration and recovery of brain serotonin neurons, *Pharmacol. Biochem. Behav.* 29(2), 269–274, 1988.

74. Scanzello, C.R., Hatzidimitriou, G., Martello, A.L., Katz, J.L., and Ricaurte, G.A., Serotonergic recovery after (+/–)3,4-(methylenedioxy) methamphetamine injury: observations in rats, *J. Pharmacol. Exp. Ther.* 264(3), 1484–1491, 1993.

75. Sprague, J.E., Everman, S.L., and Nichols, D.E., An integrated hypothesis for the serotonergic axonal loss induced by 3,4-methylenedioxymethamphetamine, *Neurotoxicology* 19(3), 427–441, 1998.

76. Kuhn, D.M. and Geddes, T.J., Molecular footprints of neurotoxic amphetamine action, *Ann. N.Y. Acad. Sci.* 914, 92–103, 2000.

77. Gudelsky, G.A. and Yamamoto, B.K., Neuropharmacology and neurotoxicity of 3,4-methylene-dioxymethamphetamine, *Methods Mol. Med.* 79, 55–73, 2003.

78. Malberg, J.E. and Seiden, L.S., Small changes in ambient temperature cause large changes in 3,4-methylenedioxymethamphetamine (MDMA)-induced serotonin neurotoxicity and core body temper-ature in the rat, *J. Neurosci.* 18(13), 5086–5094, 1998.

79. Green, A.R., O'Shea, E., and Colado, M.I., A review of the mechanisms involved in the acute MDMA (ecstasy)-induced hyperthermic response, *Eur. J. Pharmacol.* 500(1–3), 3–13, 2004.

80. Carlsson, A., The contribution of drug research to investigating the nature of endogenous depression, *Pharmakopsychiatr. Neuropsychopharmakol.* 9(1), 2–10, 1976.

81. Benmansour, S., Cecchi, M., Morilak, D.A., Gerhardt, G.A., Javors, M.A., Gould, G.G., and Frazer, A., Effects of chronic antidepressant treatments on serotonin transporter function, density, and mRNA level, *J. Neurosci.* 19(23), 10494–10501, 1999.

82. Benmansour, S., Owens, W.A., Cecchi, M., Morilak, D.A., and Frazer, A., Serotonin clearance *in vivo* is altered to a greater extent by antidepressant-induced downregulation of the serotonin transporter than by acute blockade of this transporter, *J. Neurosci.* 22(15), 6766–6772, 2002.

83. Frazer, A. and Benmansour, S., Delayed pharmacological effects of antidepressants, *Mol. Psychiatry* 7(Suppl. 1), S23–S28, 2002.

84. Kalia, M., O'Callaghan, J.P., Miller, D.B., and Kramer, M., Comparative study of fluoxetine, sibutra-mine, sertraline and dexfenfluramine on the morphology of serotonergic nerve terminals using sero-tonin immunohistochemistry, *Brain Res.* 858(1), 92–105, 2000.

85. Gouzoulis-Mayfrank, E., Becker, S., Pelz, S., Tuchtenhagen, F., and Daumann, J., Neuroendocrine abnormalities in recreational ecstasy (MDMA) users: is it ecstasy or cannabis? *Biol. Psychiatry* 51(9), 766–769, 2002.

86. Kish, S.J., How strong is the evidence that brain serotonin neurons are damaged in human users of ecstasy? *Pharmacol. Biochem. Behav.* 71(4), 845–855, 2002.

87. Reneman, L., Designer drugs: how dangerous are they? *J. Neural Transm. Suppl.* 66, 61–83, 2003.

88. Gartside, S.E., McQuade, R., and Sharp, T., Effects of repeated administration of 3,4-methylene-dioxymethamphetamine on 5-hydroxytryptamine neuronal activity and release in the rat brain *in vivo*, *J. Pharmacol. Exp. Ther.* 279(1), 277–283, 1996.

89. Aghajanian, G.K., Wang, R.Y., and Baraban, J., Serotonergic and non-serotonergic neurons of the dorsal raphe: reciprocal changes in firing induced by peripheral nerve stimulation, *Brain Res.* 153(1), 169–175, 1978.

90. Hajos, M. and Sharp, T., A 5-hydroxytryptamine lesion markedly reduces the incidence of burst-firing dorsal raphe neurones in the rat, *Neurosci. Lett.* 204(3), 161–164, 1996.

91. Switzer, R.C., III, Application of silver degeneration stains for neurotoxicity testing, *Toxicol. Pathol.* 28(1), 70–83, 2000.

92. Jensen, K.F., Olin, J., Haykal-Coates, N., O'Callaghan, J., Miller, D.B., and de Olmos, J.S., Mapping toxicant-induced nervous system damage with a cupric silver stain: a quantitative analysis of neural degeneration induced by 3,4-methylenedioxymethamphetamine, *NIDA Res. Monogr.* 136, 133–149; discussion 150–154, 1993.

93. Steinbusch, H.W., Distribution of serotonin-immunoreactivity in the central nervous system of the rat-cell bodies and terminals, *Neuroscience* 6(4), 557–618, 1981.

94. O'Callaghan, J.P. and Sriram, K., Glial fibrillary acidic protein and related glial proteins as biomarkers of neurotoxicity, *Expert Opin. Drug Saf.* 4(3), 433–442, 2005.

95. O'Callaghan, J.P., Jensen, K.F., and Miller, D.B., Quantitative aspects of drug and toxicant-induced astrogliosis, *Neurochem. Int.* 26(2), 115–124, 1995.

96. O'Callaghan, J.P. and Miller, D.B., Quantification of reactive gliosis as an approach to neurotoxicity assessment, *NIDA Res. Monogr.* 136, 188–212, 1993.

97. Wang, X., Baumann, M.H., Xu, H., and Rothman, R.B., 3,4-Methylenedioxymethamphetamine (MDMA) administration to rats decreases brain tissue serotonin but not serotonin transporter protein and glial fibrillary acidic protein, *Synapse* 53(4), 240–248, 2004.

98. Pubill, D., Canudas, A.M., Pallas, M., Camins, A., Camarasa, J., and Escubedo, E., Different glial response to methamphetamine- and methylenedioxymethamphetamine-induced neurotoxicity, *Naunyn Schmiedeberg's Arch. Pharmacol.* 367(5), 490–499, 2003.

99. Wang, X., Baumann, M.H., Xu, H., Morales, M., and Rothman, R.B., ({+/–})-3,4-Methylene-dioxymethamphetamine administration to rats does not decrease levels of the serotonin transporter protein or alter its distribution between endosomes and the plasma membrane, *J. Pharmacol. Exp. Ther.* 314(3), 1002–1012, 2005.

100. Ricaurte, G.A., Yuan, J., and McCann, U.D., (+/–)3,4-Methylenedioxymethamphetamine ("Ecstasy")-induced serotonin neurotoxicity: studies in animals, *Neuropsychobiology* 42(1), 5–10, 2000.

101. White, C.R. and Seymour, R.S., Allometric scaling of mammalian metabolism, *J. Exp. Biol.* 208(9), 1611–1619, 2005.

102. West, G.B. and Brown, J.H., The origin of allometric scaling laws in biology from genomes to ecosystems: towards a quantitative unifying theory of biological structure and organization, *J. Exp. Biol.* 208(9), 1575–1592, 2005.

103. Lin, J.H., Applications and limitations of interspecies scaling and *in vitro* extrapolation in pharmaco-kinetics, *Drug Metab. Dispos.* 26(12), 1202–1212, 1998.

104. Mahmood, I., Allometric issues in drug development, *J. Pharm. Sci.* 88(11), 1101–1106, 1999.

105. Campbell, D.B., The use of toxicokinetics for the safety assessment of drugs acting in the brain, *Mol. Neurobiol.* 11(1–3), 193–216, 1995.

106. Liechti, M.E. and Vollenweider, F.X., The serotonin uptake inhibitor citalopram reduces acute car-diovascular and vegetative effects of 3,4-methylenedioxymethamphetamine ("Ecstasy") in healthy volunteers, *J. Psychopharmacol.* 14(3), 269–274, 2000.

107. Glennon, R.A. and Higgs, R., Investigation of MDMA-related agents in rats trained to discriminate MDMA from saline, *Pharmacol. Biochem. Behav.* 43(3), 759–763, 1992.

108. Schechter, M.D., Serotonergic-dopaminergic mediation of 3,4-methylenedioxymethamphetamine (MDMA, "ecstasy"), *Pharmacol. Biochem. Behav.* 31(4), 817–824, 1988.

109. Johanson, C.E., Kilbey, M., Gatchalian, K., and Tancer, M., Discriminative stimulus effects of 3,4-methylenedioxymethamphetamine (MDMA) in humans trained to discriminate among *d*-amphet-amine, meta-chlorophenylpiperazine and placebo, *Drug Alcohol Depend.* 81, 27–36, 2006.

110. Schenk, S., Gittings, D., Johnstone, M., and Daniela, E., Development, maintenance and temporal pattern of self-administration maintained by ecstasy (MDMA) in rats, *Psychopharmacology* (Berlin) 169(1), 21–27, 2003.

111. Tancer, M. and Johanson, C.E., Reinforcing, subjective, and physiological effects of MDMA in humans: a comparison with *d*-amphetamine and mCPP, *Drug Alcohol Depend.* 72(1), 33–44, 2003.

112. Baumann, M.H., Ayestas, M.A., Dersch, C.M., and Rothman, R.B., 1-(*m*-Chlorophenyl)piperazine (mCPP) dissociates *in vivo* serotonin release from long-term serotonin depletion in rat brain, *Neuro-psychopharmacology* 24(5), 492–501, 2001.

113. O'Shea, E., Granados, R., Esteban, B., Colado, M.I., and Green, A.R., The relationship between the degree of neurodegeneration of rat brain 5-HT nerve terminals and the dose and frequency of administration of MDMA ("ecstasy"), *Neuropharmacology* 37(7), 919–926, 1998.

114. Series, H.G., Cowen, P.J., and Sharp, T., *p*-Chloroamphetamine (PCA), 3,4-methylenedioxy-metham-phetamine (MDMA) and *d*-fenfluramine pretreatment attenuates *d*-fenfluramine-evoked release of 5-HT *in vivo*, *Psychopharmacology* (Berlin) 116(4), 508–514, 1994.

115. Shankaran, M. and Gudelsky, G.A., A neurotoxic regimen of MDMA suppresses behavioral, thermal and neurochemical responses to subsequent MDMA administration, *Psychopharmacology* (Berlin) 147(1), 66–72, 1999.

116. Matuszewich, L., Filon, M.E., Finn, D.A., and Yamamoto, B.K., Altered forebrain neurotransmitter responses to immobilization stress following 3,4-methylenedioxymethamphetamine, *Neuroscience* 110(1), 41–48, 2002.

117. Kirby, L.G., Kreiss, D.S., Singh, A., and Lucki, I., Effect of destruction of serotonin neurons on basal and fenfluramine-induced serotonin release in striatum, *Synapse* 20(2), 99–105, 1995.

118. Romero, L., Jernej, B., Bel, N., Cicin-Sain, L., Cortes, R., and Artigas, F., Basal and stimulated extracellular serotonin concentration in the brain of rats with altered serotonin uptake, *Synapse* 28(4), 313–321, 1998.

119. Hall, F.S., Devries, A.C., Fong, G.W., Huang, S., and Pert, A., Effects of 5,7-dihydroxytryptamine depletion of tissue serotonin levels on extracellular serotonin in the striatum assessed with *in vivo* microdialysis: relationship to behavior, *Synapse* 33(1), 16–25, 1999.

120. Zigmond, M.J., Abercrombie, E.D., Berger, T.W., Grace, A.A., and Stricker, E.M., Compensations after lesions of central dopaminergic neurons: some clinical and basic implications, *Trends Neurosci.* 13(7), 290–296, 1990.

121. Van de Kar, L.D., Neuroendocrine pharmacology of serotonergic (5-HT) neurons, *Annu. Rev. Pharmacol. Toxicol.* 31, 289–320, 1991.

122. Levy, A.D., Baumann, M.H., and Van de Kar, L.D., Monoaminergic regulation of neuroendocrine function and its modification by cocaine, *Front. Neuroendocrinol.* 15(2), 85–156, 1994.

123. Series, H.G., le Masurier, M., Gartside, S.E., Franklin, M., and Sharp, T., Behavioural and neuroendocrine responses to *d*-fenfluramine in rats treated with neurotoxic amphetamines, *J. Psychopharmacol.* 9, 214–222, 1995.

124. Poland, R.E., Diminished corticotropin and enhanced prolactin responses to 8-hydroxy-2(di-*n*-propylamino)tetralin in methylenedioxymethamphetamine pretreated rats, *Neuropharmacology* 29(11), 1099–1101, 1990.

125. Poland, R.E., Lutchmansingh, P., McCracken, J.T., Zhao, J.P., Brammer, G.L., Grob, C.S., Boone, K.B., and Pechnick, R.N., Abnormal ACTH and prolactin responses to fenfluramine in rats exposed to single and multiple doses of MDMA, *Psychopharmacology* (Berlin) 131(4), 411–419, 1997.

126. Baumann, M.H., Ayestas, M.A., and Rothman, R.B., Functional consequences of central serotonin depletion produced by repeated fenfluramine administration in rats, *J. Neurosci.* 18(21), 9069–9077, 1998.

127. Gerra, G., Zaimovic, A., Giucastro, G., Maestri, D., Monica, C., Sartori, R., Caccavari, R., and Delsignore, R., Serotonergic function after (+/−)3,4-methylene-dioxymethamphetamine ("Ecstasy") in humans, *Int. Clin. Psychopharmacol.* 13(1), 1–9, 1998.

128. Gerra, G., Zaimovic, A., Ferri, M., Zambelli, U., Timpano, M., Neri, E., Marzocchi, G.F., Delsignore, R., and Brambilla, F., Long-lasting effects of (+/−)3,4-methylenedioxymethamphetamine (ecstasy) on serotonin system function in humans, *Biol. Psychiatry* 47(2), 127–136, 2000.

129. Seiden, L.S., Woolverton, W.L., Lorens, S.A., Williams, J.E., Corwin, R.L., Hata, N., and Olimski, M., Behavioral consequences of partial monoamine depletion in the CNS after methamphetamine-like drugs: the conflict between pharmacology and toxicology, *NIDA Res. Monogr.* 136, 34–46; discussion 46–52, 1993.

130. Robinson, T.E., Castaneda, E., and Whishaw, I.Q., Effects of cortical serotonin depletion induced by 3,4-methylenedioxymethamphetamine (MDMA) on behavior, before and after additional cholinergic blockade, *Neuropsychopharmacology* 8(1), 77–85, 1993.

131. Ricaurte, G.A., Markowska, A.L., Wenk, G.L., Hatzidimitriou, G., Wlos, J., and Olton, D.S., 3,4-Methylenedioxymethamphetamine, serotonin and memory, *J. Pharmacol. Exp. Ther.* 266(2), 1097–1105, 1993.

132. McNamara, M.G., Kelly, J.P., and Leonard, B.E., Some behavioural and neurochemical aspects of subacute (+/−)3,4-methylenedioxymethamphetamine administration in rats, *Pharmacol. Biochem. Behav.* 52(3), 479–484, 1995.

133. Byrne, T., Baker, L.E., and Poling, A., MDMA and learning: effects of acute and neurotoxic exposure in the rat, *Pharmacol. Biochem. Behav.* 66(3), 501–508, 2000.

134. Marston, H.M., Reid, M.E., Lawrence, J.A., Olverman, H.J., and Butcher, S.P., Behavioural analysis of the acute and chronic effects of MDMA treatment in the rat, *Psychopharmacology* (Berlin) 144(1), 67–76, 1999.

135. Morley, K.C., Gallate, J.E., Hunt, G.E., Mallet, P.E., and McGregor, I.S., Increased anxiety and impaired memory in rats 3 months after administration of 3,4-methylenedioxymethamphetamine ("ecstasy"), *Eur. J. Pharmacol.* 433(1), 91–99, 2001.

136. Gurtman, C.G., Morley, K.C., Li, K.M., Hunt, G.E., and McGregor, I.S., Increased anxiety in rats after 3,4-methylenedioxymethamphetamine: association with serotonin depletion, *Eur. J. Pharmacol.* 446(1–3), 89–96, 2002.

137. Fone, K.C., Beckett, S.R., Topham, I.A., Swettenham, J., Ball, M., and Maddocks, L., Long-term changes in social interaction and reward following repeated MDMA administration to adolescent rats without accompanying serotonergic neurotoxicity, *Psychopharmacology* (Berlin) 159(4), 437–444, 2002.

138. McGregor, I.S., Clemens, K.J., Van der Plasse, G., Li, K.M., Hunt, G.E., Chen, F., and Lawrence, A.J., Increased anxiety 3 months after brief exposure to MDMA ("Ecstasy") in rats: association with altered 5-HT transporter and receptor density, *Neuropsychopharmacology* 28(8), 1472–1484, 2003.

139. Bull, E.J., Hutson, P.H., and Fone, K.C., Reduced social interaction following 3,4-methylene-dioxymethamphetamine is not associated with enhanced 5-HT 2C receptor responsivity, *Neuropharmacology* 44(4), 439–448, 2003.

140. Bull, E.J., Hutson, P.H., and Fone, K.C., Decreased social behaviour following 3,4-methylene-dioxymethamphetamine (MDMA) is accompanied by changes in 5-HT2A receptor responsivity, *Neuropharmacology* 46(2), 202–210, 2004.

Index

W

Z